Collecting Crystals

The Guide to Arkansas Quartz

J Michael Howard

A&I Studio Press
Rockhounding Arkansas

Collecting Crystals
The Guide to Arkansas Quartz
Copyright © 2000 by Darcy and Mike Howard

Illustrations by Darcy Howard
Photographs are by the authors
All rights reserved

ISBN 0-9677300-0-7

Library of Congress Catalog Card Number: 99-068340

Manufactured in the United States of America by Sheridan Books, Inc.

Except as permitted under the United States Copyright Act of 1976, no part of this publication may be reproduced or distributed in any form or by any means, or stored in a data base or retrieval system, without the prior written permission of the publisher.

A&I Studio Press • P.O. Box 45791 • Little Rock AR 72214 • USA • 501-847-6575

Rockhounding Arkansas   www.rockhoundingAR.com

Front cover, quartz crystal from Ron Coleman mine, back cover, quartz crystal from Mount Ida region.

Acknowledgments

A special thanks to the people who helped make this book possible: Henry de Linde, Miriam Talbert, Art Smith, Don Owens of the UALR Earth Science Department, and John Nichols of the US Forest Service for their reviews. To David and Sara Dodson for their experience of rockhounding, computers and printing, and to the shop and mine owners who took time to visit with us.

Dedicated to those who found pebbles in the stream as a child and have been picking up stones ever since.

Contents

Introduction 1

1. **Crystals in Arkansas**............................... 5
   The major crystal belt
2. **Getting Ready to Collect** 19
   Tools and equipment you will need
3. **Collecting the Crystals** 33
   Surface collecting and mining the veins
4. **Cleaning and Trimming** 45
   Making your treasures look their best
5. **Places to Collect** 63
   Public access free areas, and fee pay mines
6. **Kinds of Crystals** 89
   Varieties of collectable quartz
7. **Other Quartz in Arkansas**..................... 107
   Found outside the major belt
8. **Mineralogy of Quartz** 115
   Physical properties of the mineral
9. **Frequently Asked Questions**.................. 129
   Answers to common concerns
10. **Facilities and Resources**........................ 143
    Campsites, lodging, and places to get more information

Maps of collecting areas...................... 85 to 87

Glossary.. 153

Introduction

MANY collectors relate their first experience picking up rocks was finding smooth, shiny pebbles in a creek. Some of these people have gone on to complete a Ph.D. in geology, and others, like me, have a few rocks stuck in a drawer. But between these two extremes lie the majority who found a quartz crystal and have been struck with its beauty, leading to a quest for finding more of these fascinating "frozen ices" of nature.

The rockhounding bug, or rock pox as some call it, sometimes affects children, but adults are usually the ones who can't resist the call of an irresistible stone. Crystals have a way of asking to be picked up. One pretty crystal asks for another, and soon, a box is filled and another collector is taken over by these beautiful clear natural jewels.

Living in central Arkansas, we meet many collectors who travel to the area just to dig crystals. Most of these people also collect other minerals, and a good number tell me it was quartz crystal that really got them interested in the hobby.

My husband fell into the clutches of collecting quartz when he was just eight years old. Now, many years later, he is still as enthusiastic about going on a crystal dig as he was in his younger years. Let me explain why we decided to write this book. Mike is a mineral collector for a hobby, and a geologist for a living. He writes and teaches the science of mineralogy. He collects and brings home crystals and he knows a tremendous amount about crystals. We've had crystals in our kitchen sink, bathroom sink, they've gone through the washing machine in his pockets, every horizontal surface in our home has had crystals on it at one time or another. Crystal display cabinets take up more room than our entertainment center. So you get the idea he *really* likes crystals...

His collector friends always like to ask him questions, and we get a lot of questions from our website. Having learned a bit of mineralogy by marriage, my job here is to translate the often-complex answers he gives these people into concepts the rest of us can understand. And as for picking brains, while researching this book we drove for hours and miles, travelling from one mine to another, Mike talking and me scribbling notes. Most of the work in this book is Mike's experience, and my observations of how other collectors can benefit from the information an enthusiastic professional brings to the hobby of mineral collecting.

As we've traveled the state and country we've met and made good friends. Many coming here to dig crystals and visit have become close friends of ours. We would never have had a chance to know so many from outside the United States without the mineral connection.

We enjoy getting e-mail from our website and from other collectors who tell us their stories, and we've met intrepid souls who have been collecting for as long as seventy-five years! Some of the students in our Elderhostel classes go on their very first crystal digs with us after they have retired from their careers. Youngsters in our Cub Scout pack have gone with us and found nice crystals and clusters. Young and old alike, age doesn't matter, you can always enjoy the thrill of the hunt, the joy of finding, and the satisfaction of collecting a crystal yourself.

Families enjoy collecting crystals together, too. As a vacation activity, a crystal digging trip can easily be combined with travel, sightseeing, camping, boating, and fishing, or the many other opportunities of the central Arkansas region. Tourism in the Hot Springs National Park area is oriented toward families, and the large clean lakes of central Arkansas are an excellent area to establish a home base when you are visiting the Ouachita Mountains.

By the way, Ouachita is pronounced "WASH-i-taw". It's a French spelling of a Native American word. The locals will know you're a tourist if they hear you say the strange pronunciations phonics might prompt you to try.

When you plan your trip to collect crystals, whether for an afternoon, a weekend, or a week, leave yourself some time to look around. With a national park, several lakes, and many acres of National Forest land in the same region as the quartz, you'll find an abundant choice of recreational opportunities.

We both write in this book in the first person. Together we'll tell you about the best places to collect, and how to make the best of your rockhounding trip. If you've never dug crystals, you are in for a fascinating time, and if you are an experienced collector, you will find information in this book to enhance your enjoyment of the hobby.

Good luck and happy digging!

Darcy Howard
Rockhounding Arkansas
ARrockinfo@aol.com

1. Crystals in Arkansas

THE OLD TIMERS in this land called the clear quartz points "rock crystal". Long before our grandfathers were picking up rock crystal, the Native Americans were using crystal for special purposes. For as many generations as people have been in the land we now call Arkansas, there have been collectors of crystal. Crystal arrowheads have been discovered that archeologists date back to 8,000 BC.

The Plum Bayou people, an unknown tribe of uncertain ancestry, built mounds near the Arkansas River around 700 to 1000 AD. These mounds were not burial mounds, but are believed to have had astronomical or other ceremonial purposes. Projectile points, scrapers and knives associated with these people were found crafted out of clear quartz. Because crystal is so much more difficult to work with than other locally available stone, archeologists believe these people must have had a determined use in mind for such crystal implements, perhaps ceremonially important.

The Spanish explorer Hernando DeSoto visited Arkansas in the year 1541 looking for the legendary fountain of youth. He came to Hot Springs, didn't find what he was hoping for, but instead found a creek where hot water flowed out of the mountainside. Here the peoples of many

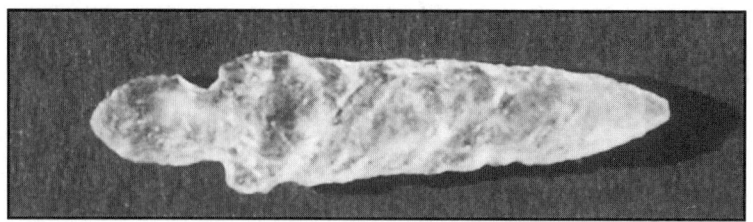

Hayes arrowhead point made from quartz crystal, dated 800-1200 AD. Found in Little River County, this piece is about two inches long and was part of a cache of about 70 quartz crystal points. From the collection of Sam Johnson, Murfreesboro, AR.

tribes came and laid down their weapons and brought their sick and injured for healing. Translations of his journals vary, but he reported the people of the area were making arrowheads from quartz crystal at that time.

Today, just like all the people before us, we're still picking up and using crystals. Each generation finds them beautifully new and fascinating. We've found different ways to use them as our technology has expanded. The applications for quartz in our society vary widely from personal enjoyment to computer, satellite, and communication applications. Because of its collectability and economic importance, our state legislature has even designated quartz as the Arkansas State Mineral.

Known around the world

There are several places in the world where quartz crystals occur in significant quantities, like Brazil, Columbia and Madagascar. The Ouachita Mountain region of Arkansas is also one of these major locations. Based on quality, quantity and the number of mines, the quartz fields in Arkansas make it one of the largest producing areas in the world. Arkansas quartz is known around the world by both collectors and the scientific community, so you've got a good reason to come here to collect!

Quartz, or silica (SiO_2), is a hard, clear and durable mineral that is greatly resistant to weathering. That's why we can find crystals laying on the ground, because they have weathered out of the surrounding rock. It comes in many varieties, but Arkansas is known mostly for the individual crystal points and clusters, although other forms do occur here, making it all the more collectable.

How quartz formed

A survey of literature and stories about quartz will yield many explanations of the origin of crystals, from being placed by the people of the lost continent of Atlantis to erupting from volcanos. In addition to—and sometimes despite of—myriads of explanations from miners, shop owners and old timers, including those with mystical knowledge, the process of crystal formation has been well studied scientifically and is understood to the degree that artificial crystals can be grown by industry. The processes of nature are so majestic that little human embellishment is needed to realize the magnificence of the earth and universe itself.

There are two mechanisms that can cause quartz to form, one is related to magmatic activity and the other is from hot water solutions saturated with large amounts of dissolved silica from deep in the earth. Arkansas quartz is from the second category. In the next few pages we'll take you back over the eons in a time-machine fashion and tell you how the crystals came to be.

Looking at a geologic map on page 9, we find that most of the quartz veins and crystal deposits occur in the heart of the Ouachita mountains, in a belt 30 to 40 miles wide that stretches across western Arkansas into eastern Oklahoma. We refer often to this belt of quartz in later chapters. The most productive quartz veins are present here in rocks

The Ouachitas near Mount Ida provide timber, water, rangeland and recreational opportunities.

called sandstones and shales that are Paleozoic in age.

The crystal story is tied to the Ouachita Mountains, which have a long and complex history. A little background information about these mountains will show you how the quartz crystals got into the area. Geologic time spans so many human lifetimes that it is hard for those who have not studied geology to understand just how many years it has taken for the Ouachita Mountains to become what they are today.

Geologic history

Mountains can be built in several ways: by volcanoes, by a block of land that is squeezed up (uplifted) between two faults in the earth's crust, by erosion—as seen in North Arkansas, or by folding and faulting where tectonic plates pushed together. The Ouachita Mountains are folded and faulted mountains.

In a time long, long before the dinosaurs, when coal-

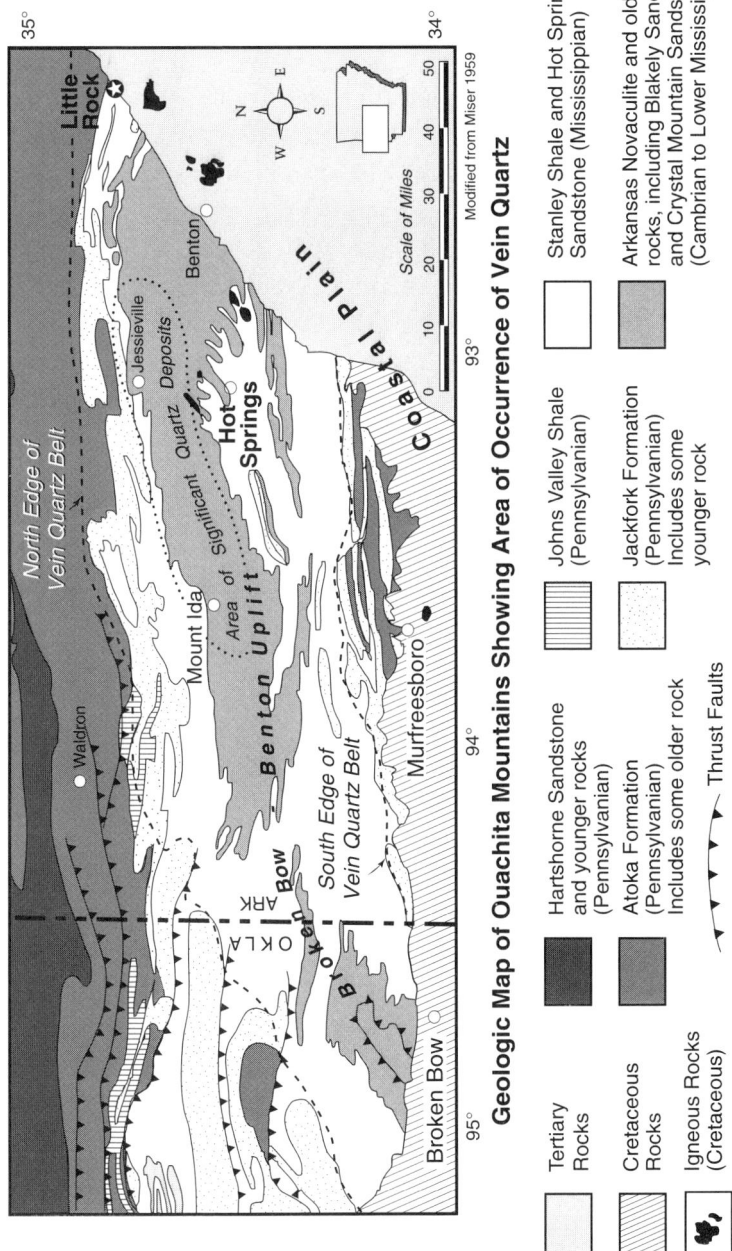

Geologic Map of Ouachita Mountains Showing Area of Occurrence of Vein Quartz

forming forests of North America flourished, the area that is now Arkansas was on the shore of an ancient ocean. The continents weren't shaped quite the same as they are now, and they weren't even in the same places as they are today. The theory of plate tectonics says that the continents of the world ride on slowly moving sections of the earth's crust, conveyor-belt style, and the plates and continents bump into and slide past each other. Bringing up new material from rift valleys and spreading sea floors, volcanoes are a feature of plate margins where the earth is cracking apart. In other places, plate movements go down into the earth, causing troughs. Plates like California slide past each other. Some places the continents bumped together, as in the Ouachitas! Keep these thoughts in mind for a couple of paragraphs.

The ancient North American continent had the same forces of erosion working on it then as we have now. Just like the the muddy Mississippi River carries sediment to the Gulf of Mexico, ancestral rivers carried sediment for eons into the ocean, causing layers of mud to build up over millions of years and compact into rocks. The transportation processes of the rivers and the sea sorted the sediments into sands and clays, later to become sandstones and shales. Sea animals lived and died and were buried in these layers, creating a fossil record of the time they were alive. There are not many fossils in the Ouachita Mountains, but some, such as graptolites—which look remarkably like jigsaw blades—can be seen in the rocks around Lake Ouachita. These many layers of sediment collected in the deep part of an old ocean that geologists call the Ouachita Basin, where central and southern Arkansas would someday be.

Changes were taking place over these ages as the silt and mud were accumulating. The movements of the continents was continuously going on, as they still are today. Geologists think South America was inching this way, making the ocean area smaller as the plates converged. The ocean floor rippled up against the continental margin of North America. The floor of the ocean rose, making the ocean shallower as South America came closer. The sediments, which were now sedimentary rock, were squeezed and folded like a rug that gets pushed against a wall. In a slow motion crash the other continent kept pushing against these layers of sediment with mountain-building power and forced them up out of the ocean, welding new mountains to the continental shelf of North America. In places, large sheets of land were thrust northward, overriding younger rock units. Brittle rocks of these sheets cracked and faulted from the buildup of pressure.

Quartz began to form

So pressures that built up in the rock layers caused strain that ruptured and cracked the hard sandstone units. In the meantime, hot silica-rich fluids seeped into those cracks in the deeper rocks inside this region. Quartz began to form on

Veins of white quartz cut through massive grey sandstone of the Crystal Mountain Formation.

the sides of the open fractures, a little like the rock candy we can grow from hot sugar syrup at home in our kitchens. Those new mountains of folded, cracked, and faulted beds of sedimentary rock were as high then as the Rocky Mountains are today. It took tens of millions of years for mountain building to be completed. By the time the dinosaurs arrived on the scene, some 150 million years after the mountains building, Arkansas had a set of mountains with quartz crystals deep inside them, and a shallow Gulf of Mexico covering its southern lands.

Erosion has been working on the mountains ever since they first became dry land. Today the Ouachita Mountains have been reduced to a mere fraction of themselves, so that we now see rocks at the surface that were originally a mile or two deep in the ground. And that uneroded Ouachita Mountain range itself was once just sediment at the bottom of the ocean. Remarkable as it may seem, sparse fossils of long dead sea animals are preserved in the Ouachita Mountains still and are occasionally discovered.

Quartz veins in the major crystal belt

The above is a very brief summary of Arkansas geologic history over the past 300 million years. Getting back to the crystals, hot fluids with dissolved minerals, especially silica, traveled along the fractures in the rocks. Miles below the surface of the ground, the earth is much like a pressure cooker, and hot fluids from the thick layers of sediments were squeezed out. Moving upward from an area of high pressure to lower pressure, major faults were pathways for the escaping fluids. The silica-laden fluids filled open fractures, where conditions were favorable for quartz crystals to form, again, like our rock candy example.

The hydrothermal fluid source was brines originally trapped in the sedimentary units as they were deposited. They moved upwards filling open fractures in the rock. These fluids appear to have dissolved silica locally and deposited it only short distances from where it originated.

Most of the quartz in the Ouachita mountains is milky quartz, in which tiny air bubbles throughout the mineral cloud transmission of light. In his classic USGS Bulletin 973-E, A. E. J. Engels noted that "massive milky quartz …dominates in most deposits." Indeed there are huge milky quartz veins up to 60 feet wide and hundreds of feet long that are sources for industrial quartz in this belt region we're talking about.

Quartz veins formed along fractures, and can be seen weaving through the rock structures, often in complicated patterns. The quartz started being deposited on the walls of open fractures, and grew inward. Where the crystal didn't grow entirely across the opening, a hole was left, which miners call a pocket. Particularly large pockets often represent the intersections of two or more open fractures. Pockets can be a few inches to many feet across.

As the mountains settled and adjusted, more faulting and fracturing occurred, occasionally causing the veins, pockets, and crystals themselves to break. Sometimes they rehealed. Most of the quartz infilling seems to have occurred in the Late Pennsylvanian age (about 290 million years ago) as the mountains were building, and growth appears to have stopped by the Triassic period (245 million years ago), the time when early marine reptiles appeared.

Much later, weathering of overlying rock units formed a red-orange clay that filled the rest of the pockets, and today is a protective packaging for quartz crystals still in the ground. This clay can be an indicator of a quartz pocket, and sharp-eyed diggers are always on the lookout for it.

Quartz veins are most numerous along the central core of the Ouachita Mountain region (see map page 9), where they are present in shale, slate, sandstone, novaculite and chert. Some veins are as wide as 60 feet in Arkansas, and nearly 100 feet in Oklahoma. Most of the quartz mines in the major crystal belt are in the Crystal Mountain and Blakely sandstones (dated by fossils to be Ordovician in age, or 500 to 440 million years old). These units have a high proportion of clear crystals in cavities or pockets.

Two major areas

Presently there are two major areas of quartz crystal mining, the Mount Ida area in Montgomery County, and the Jessieville area in northeastern Garland County. The Mount Ida area generally yields rather small and clear single crystal points plus some sizeable clusters from the Crystal Mountain sandstone. The Jessieville location is known for large, clear to milky quartz crystals and clusters which come from the Blakely sandstone. Singles and clusters from both locations are beautiful. Pieces from each location have a slightly different character and appearance.

We find the Mount Ida clusters to be more delicately beautiful, with perhaps a little greater clarity and gray cast, while the Jessieville crystals are milkier but boldly impressive because of their size. A point well worth mentioning is that a crystal or cluster doesn't have to be big to be beautiful or expensive, and small ones can be museum quality, too.

The mines then and now

As we toured through the Crystal Mountains area of Mount Ida, I tried to get a feel for the names of all the mines. It appears as leases are taken and relinquished, and new holes are dug, the names of the mines change according to who is working the pit and digging a new excavation. Miners want to call each hole a different name, even though it is on the same property. For example, Tim Hill's property also has another mine called the Drain Hole, which sounded all the world to me like a movie showing how to become a starving miner. Not so, it turned out, for the pit was christened for a previous miner named Coy Drain. So it is that the land may be owned and operated by the owner, or leased, or any sort of arrangement, which can make it hard to know who to ask for permission to collect. The National Forest Service can come to the rescue with property maps, so in case of doubt, ask the district ranger.

Many claims have come and gone over the years, but the fee pay mines seem to be stable. You'll hear names like Fisher Mountain, Collier Creek, High Peak, Miller Mountain, Coleman mine and others. These places have a long history of crystal mining.

Family connections to crystals are strong. In the Jessieville area, Charlie Coleman began mining crystal in 1946 with a pick and shovel. Today his sons Ron and Jim operate the largest quartz mines in the state. Mount Ida likewise has generations of crystal miners. In World War II, Will Fisher dug oscillator quartz for the government. One day he asked his son-in-law Ocus Stanley to help take out a large load with a horse and wagon. Ocus was so struck with the crystal he was hooked for life, and was later dubbed the Grand Old Man of Quartz by a well-known USGS geologist, Hugh Miser. The Stanley home place and crystal yard is in the same location as it was in the 1940's, now occupied by Sonny Stanley.

The crystal market

Significant sales of crystals have been reported since the 1800's when settlers came to Arkansas. When the white man started mining crystal, it was mostly a farmer's occupation, done in the off season to find crystals for tourists in the nearby international resort town of Hot Springs. Radio oscillators became an important commodity in World War II, and that's when finding clear crystal became serious business. During the war, the government war board considered quartz essential. Nationwide, all other mining not considered essential to the war effort had ceased. Using picks and shovels, miners went to work on the Dierks Timber and Coal Company land (now the Ron Coleman mine), and in the Crystal Mountains of Mount Ida. If they could get one pound of good, clear oscillator grade quartz, they could sell it for $60, roughly equal to one month's wages in those years. This period of time resulted in the only underground adits dug in the Ouachitas for quartz, these mines typically being open pit operations.

Times changed, and up until the 1980's, stories go that many a pocket of quartz was sold for a case of beer. In the early 1980's, the price of the best quality quartz was about $30 per pound.

About the same time, several popular books touted quartz crystal as New Age adjuncts to reaching a cosmic consciousness. The notion quickly generated a market driven demand for specimen grade quartz crystal, causing the price of attractive single points and clusters to escalate to an all time high. Also, raw material for industrial needs of electronic equipment brought another economic opportunity for quartz producers. Fueled by the prospects of making fast dollars, many entrepreneurs set themselves up in the quartz mining business.

The quartz boom lasted only a few years. Foreign market input to quartz crystal made prices fall, and many Arkansas miners were forced out of business just a short time later.

Up until the middle 1980's, the market for quartz was the specimen collector, tourists, and museums. Today, at the turn of the century, New Age demand for crystal is a large part of the market, and dealers who themselves may not believe in the mystical powers of quartz have learned to identify the markings and characteristics of certain crystals in order to sell them for "healing".

At the end of the 1990's, prices of quartz varied from dealer to dealer, with jewelry points, specimen grade material and metaphysical quartz bringing the highest prices. Very good specimen crystal could be purchased for $50 per pound, while jewelry points were selling from $100 to $200 per pound.

A survey of dealers showed differing prices. Since there is no absolute pricing standards for quartz, I inquired from several dealers what prices are based on. There are several factors to consider, mentioned in a later chapter, but we had to chuckle as one man said something to the effect of "Well, if the electric bill is due..."

And they ran up hill and down dale, knapping the chunky stones to pieces with hammers, like so many road makers gone daft.

They say it is to see how the world was made.

Sir Walter Scott

2. Getting Ready to Collect

THE RIGHT kinds of tools will make your job easier when you are collecting crystals. You need to select tools depending on where you are going, what kind of collecting you are going to do, and how long you can stay at the site.

If you are about to go on your first quartz digging trip, don't be alarmed by the list of tools we talk about in this chapter. Your eyes and your fingers are your only essential requirements, everything else just makes collecting more productive. You don't have to go out and buy all of the items listed here, but serious collectors will have accumulated most, if not all, of these tools.

If you have only a short time to look around a place, say an hour, you'll probably just want to look over the area and take a small garden tool for surface scratching, or use the pick end of a rock hammer. If you are going to a fee-pay mine, some owners have tools they will either loan or rent. The conditions of fee-pay mines vary as to the areas you work in, and are described in Chapter 5. That section will be useful in planning your trip.

If you are going for an afternoon, which is long enough to wear out most of us, you can justify taking a few more

Checklist

Equipment
- hand lens
- garden scratcher
- rock hammer
- small sledge
- big sledge
- pry bar
- shovel
- chisels
- whisk broom
- old towel

Personal Gear
- gloves
- sturdy footwear
- eye protection
- old clothes
- sit upon
- bug spray
- first aid kit
- sunscreen
- hat
- hardhat
- note paper and pen for writing labels
- water for drinking and washing your hands

Packing
- newspaper
- beer flats
- egg cartons
- cardboard boxes
- buckets
- canvas tote bag
- trash bag

tools. For an expedition to find the Mother Lode, that's when knowing about the right tools and having a good selection will help you the most. You can take a truckload of tools pretty easily, and I've yet to meet a rockhound who hasn't wistfully longed for a backhoe.

Here are the standard tools that go with us in the field.

Hand lens

A good hand lens, or magnifier, is an essential tool for anyone who wants to look at small crystals. When you buy one, you get what you pay for. Any in the price range from about $10 to $35 are good. A cheap one is not worth having. If you are buying a hand lens for the first time, get one with 10 power magnification (10X).

To use a hand lens, you need two things, adequate light and a clean lens! Sunlight is the best. Use a soft cotton cloth, like an old T-shirt to clean the lens. You will find that the lens needs to be equal distance from your eye and from the mineral you want to see. It takes a little practice to get used to holding the lens only an inch or so from your eye and the specimen up close, but pretty quickly you can learn to use one like a geologist. Putting a hand lens on a string around your neck

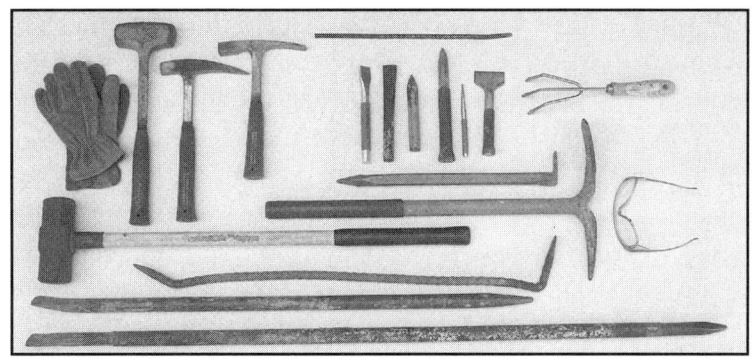

A collection of tools for digging quartz. Rock hammers on the top left, a variety of chisels in the top center, and the garden scratcher on the top right are the easier tools to work with. The hoe pick is useful for moving dirt. Pry bars with points and bends on the ends are used for getting crystals out of pockets. The big guns are the sledge hammer and the 5 foot pry bar.

makes it easy to use and quickly accessible.

Hammers

Different sizes and types of hammers are used for different purposes. They come in several varieties.

SLEDGE An 8-pound or heavier long-handle sledge is used for breaking big rocks into smaller, more manageable pieces. You have to be big and strong enough to use a sledge without hurting yourself. You can tire out quickly using a sledge hammer.

CRACK HAMMER A 4-pound or 2-pound crack hammer is handy for breaking medium-sized samples or driving chisels. If you don't have strong arms, these will tire you out, also.

GEOLOGIST'S HAMMER A geologist's hammer is useful for prying rocks apart or out of the ground because it has a pick point on one end. The flat end can be used to break rocks. The point end is made for prying, not hammering.

BRICK LAYER'S (SPLITTING) HAMMER A similar specialty hammer is called a bricklayer's hammer. Instead of a pick end, it has a chisel on one end. The brick layer's hammer may be driven, like a chisel, into rock fractures, then by using the handle as a lever, it may be used to pry out samples. It is also very handy for splitting layered sedimentary rocks or concretions.

An unusual hammer I sometimes use has a pick on one end and a chisel on the other. You may also want a small hammer for light trimming of specimens.

I recommend Estwing® products. They are expensive, but should last many years if you do not lose them (and that happens easily on an outcrop). Estwing guarantees their products against breakage, so if you break the head off you can get a replacement for free! Any good type of large sledge will do. If you buy one with a wooden handle, be sure and carry an extra handle with you to the field in case you split the original. There is another brand of geologist's hammer and sledge hammer that comes with a fiberglass handle and is comfortable to use. You can buy these hammers at rock shops or at larger hardware stores.

Very important: When using any hammer to strike rock or another steel tool, wear safety glasses for eye protection. No specimen is worth the value of an eye!

Chisels and gad points

You need a good set of these. Some are available with plastic coating on the sides. I buy Craftsman from Sears, so if they break, I can get a free replacement. You will need several sizes and lengths, depending on the conditions of the rocks you are working. Various sizes are handy for different jobs. A very long slender chisel is useful for getting way back into narrow pockets. Sometimes I can work someone's abandoned pocket and get a good specimens due to that one chisel. See the long chisel in the photo on p. 21.

Diggers are often seen working a dirt pile with a screwdriver, and if that's your tool of choice, buy one and dedicate it to digging. It might not be in too good of a condition when you get through with it.

A tip for not losing tools
When your tools get dirty from digging, they become camouflaged on the ground. Spray paint your tools with bright fluorescent paint. You might take some ribbing from your fellow collectors, but your tools won't get lost. Part of our large array of tools comes from finding what others have left behind.

Garden tools do double duty
The short-handle 3-tine scratcher is useful for moving dirt around when you are searching in a small area and in loose soil. A long handle tool will keep you from having to bend and stoop so much. Since digging a loose crystal from the dirt is much like removing a weed, any of your favorite planting or weeding devices will work when surface collecting.

SHOVELS, HOES, PICKS, ETC. A big part of collecting, other than just scratching the surface, is moving dirt. You have to be either a good dirt mover or walk over a lot of ground to find the best and most crystals.

A garden rake is a good dirt tool. We recently met a digger who cut the handles short on her garden tools so she wouldn't whap anybody in the head. She also had old socks with the toes cut off over her elbows for protection. You'll see many clever techniques in the field from experienced rockhounds. When going to an outcrop after the leaves have fallen, a leaf rake may be useful for uncovering the ground. That's not a problem at a fee-pay mine (the first thing miners do is bulldoze the trees down—and look for crystals in the roots!), but if you are going to the National Forest land, leaves are something to think about. The dense vegetation of Arkansas hinders seeing rocks and the surrounding land, so geologists often do their field work here in the late autumn and early winter. Knowing the area you are going to visit is really helpful when selecting which tools to take. In our experience, the first trip to an area is a good learning experience, and then we are better able to plan an expedition when we return. Because the tailings piles at most fee pay mines are pretty much the same, certain factors will determine the tools you'll want to take. These considerations include how energetic you are and whether the area is wet or dry from recent weather conditions. It's kind of funny watching tourists bring big picks and shovels to scratch the surface, but since you want to find crystals instead of amuse the locals, limit yourself to equipment you can handle without hurting yourself and you'll feel much better at the end of the day.

Personal gear

A word about clothing and other things. Digging is dirty work. What kind of clothing to wear? Old clothes that you can use for the trip and don't mind if they ever come clean again. The red-orange clay associated with crystals will penetrate and stain all your clothes and even your skin after a while. Shoes take the major part of the cleanup effort. A change of clothes might be advisable to bring, too. Kids (and some adults!) don't mind getting *really* dirty. A change of clothes takes care of the problem a friend of ours calls "mud butt."

Sturdy shoes or boots help make walking on uneven ground easier. Ankle support helps. Most collecting areas have roads to walk along, but to get into the good places may require going up steep slopes or getting into a hole. Other collecting areas have dirt piles to walk across or may be rutted with tracks of heavy machinery.

Eye protection is a must if you are going to be breaking rocks. If you are using a hammer to trim specimens or break rock, fragments will fly toward you as well as toward everybody around you. Safety glasses now come in decorator sunglass styles, so check with your hardware or construction supply store for these.

Gloves will save your hands from dirt, cuts and blisters. Fabric gloves will quickly wear out on your dominant hand, so either take extra ones with you, or get leather gloves. Tiny quartz crystals are like needles, and you don't want to get these in your fingers. Bricklayers do a fingertip saving trick of wrapping tape on their fingers, and I saw one resourceful quartz digger go a step even better. She wrapped the fingertips of her fabric gloves with duct tape! She assured me that she got much better mileage from her gloves after the duct tape treatment.

Hardhats are not necessary when scouring mine tailings, but are good protective gear for working near any wall where rocks may fall from above. You will likely be cautioned to stay away from highwalls at some mines, and other mines will have these areas roped or fenced off.

If you don't carry something to sit on, you are going to find the muddy, rocky ground hard on the backside. An old boat cushion or stadium cushion is ideal. Whatever you take to sit on, it will get red clay on it. Scout the yard sales for sit-upons. You really want something cheap, because it certainly will get dirty. Buckets and milk crates are good if strong enough to hold you.

If you have difficulty sitting down and getting up, a lightweight stepstool is useful for taking along. Don't let someone in your family or group be left out of a collecting adventure because they can't get around very well. It is possible to get nice crystals from just sitting in one area on a stool and scratching the ground.

Sunscreen is a necessity as few sites have shade trees

where you will want to dig. Remember about the miners bulldozing the trees? Also bug spray. Here's a note to the un-chiggered—take our advice and use repellent spray to keep these itchy red bugs off. Chiggers are the parasitic larvae of mites, and like ticks, they live in unmowed grassy areas and woods. That's not a problem in the fee-pay areas because they have well traveled roads and you are going to be in the dirt while you are digging, but the woods of the forest are prime areas for all kinds of biting bugs. There was an interesting tidbit on the radio the other morning. According to a survey by a company that makes an itch-relieving cream, based on sales of their product, the number one itchy city in the United States is Houston, Texas. The number four itchy city in the United States is tiny Hot Springs, Arkansas, and Arkansas also has the dubious honor of more cities in the itchy top 20 than any other state. Although scratching chiggers and poison ivy is one of the most satisfying activities of life, you are really should avoid these buggers in the first place when at all possible. Take a first aid kit, too. Quartz is like glass, it will cut you.

If you plan on working with heavy hand tools, remember to bring along plenty of food and drink. Sports drinks are great during the hot Arkansas summers. Broad-brimmed straw hats are good for everybody, too. Even if you are just driving around the Forest Service roads looking the places over, it's a good idea to have a canteen with you. Summertimes in Arkansas are brutally hot. Freezing water in 2 liter soda pop bottles is a handy way to pack your cooler. You'll get cold water as the ice thaws and keep your candy bars from melting at the same time. There are very few convenience stores along the paved highways, and no facilities at all on the back roads.

Take a trash bag and haul your trash out with you. This is worth putting in bold letters. More Arkansas sites have been put off limits to collectors because of rockhounds' lack of respect for property owners. Digging messes up the land enough without beer cans and trash being left behind. If you pack it in, pack it out!

Wrapping it up

After you find your crystals, you'll want to wrap your better specimens in newspaper so they won't get damaged on your trip home. If you set a cluster on top of other clusters, the tips of the crystals can be damaged. Wrap them carefully and don't put anything on top of your best pieces. You'll need boxes or buckets to pack all the rocks in. The containers need to be sturdy enough that they can support the heavy weight of the rocks. Your best points and small clusters will travel well in egg cartons.

Traditional among rock collectors is the use of beer flats, which are the cardboard box bottoms that canned beverages come in. If you meet a person loading a soft drink machine, or go to a store after the shelves have been stocked, they will be happy to let you have an armload of flats.

Sometimes it's hard to tell if kids like collecting crystals or just playing in the dirt better. Either way, they have fun!

Pebble puppies

If you are taking children with you, a garden scratching tool is ideal for them to work with. If at all possible, let them all use the same kind of digging tool. It's very easy for kids to think that finding a crystal has something to do with the power of the tool, instead of luck or having a good eye. I've heard too many times "His digger is better than mine!", so putting the youngsters on even terms to begin with might help save your sanity. A sandbox bucket with a handle is good for them to carry single points and small clusters that they find. Older children might need to have a tonnage limit on what they can bring home (adults, too!). I make this statement because some large boulders are covered with small crystal that can't be easily pried off the base rock, and kids have difficulty realizing these kinds of big rocks belong out in the yard instead of on their bookcase. It's a good idea to cover the trunk of your car with newspaper or cardboard boxes to bring home large pieces.

Name (of specimen)
Location
Collection of (your name)
Notes

Importance of labels

Even if you know where the specimen was collected, by the time you get it home with your other boxes of crystals, will you remember which place each box came from? Could other collectors guess where it came from? If you take the time to make some notes and put in each box, or label on the box itself, you'll have the information necessary for properly identifying and displaying your crystals later. Use a ballpoint pen or waterproof marker so there is no chance of the ink bleeding or washing off.

One very important thing about a good collection, private, public, scientific, or for fun, is that each piece is labeled. The information should at least state what the specimen is and where it came from. Instead of simply labeling a specimen "from Arkansas", use a specific location, such as the "Whatever crystal mine in Garland County, Arkansas". In addition, you can put other information like "collected by" and the date on the label, and even notes such as "my father went with me on this trip".

If the thought of going to a "mine" brings on claustrophobia, not to worry! All the crystal workings are out in the open, not in dark mine tunnels or caves.

Tools for working mine tailings

Again, what you need depends on the type of mine you visit. If you are working the tailings pile or mine waste dump, then you can use the following: garden scratcher, old boat cushion to sit on, several pairs of cheap gloves, a rock hammer, wrapping paper, and a bucket or cardboard boxes to put your crystals in. Add your personal gear, and it should be a small enough pile so you'll still have plenty of room in your car for the crystals you'll find.

Tools for working the veins

If you are working the veins in the ground, then you need the above tools plus a three foot pry bar with a bend on the end which is sharpened to a point, a 4-pound sledge hammer, an 8-pound sledge hammer, a 5 to 6 foot chisel-edged stout pry or breaker bar, heavy leather gloves, and a lot of energy. Your personal protection gear becomes even more important when using these heavier tools and working in those places where you require the big tools, so make sure all the members of your team have hard hats and safety glasses.

At the mines, the veins will have been exposed by heavy equipment, but the pockets and veins are worked by hand to avoid damaging the crystals. The size of veins can range from a few inches to several feet in width. One digger told me she used a deer antler in a pocket to work out the crystals, and it wouldn't damage the points. However, her partner said that antlers were for beginners, and people who want results use steel! But that's another story, and we'll talk about it in the next chapter.

The buddy system is an especially good way to go when working a difficult area, because you can watch out for each other, and take turns doing the heavy work. You can also share sympathy with each other the next day for the sore muscles you'll get!

Others in the woods

Some folks have an unfavorable reputation and are loosely called nuisance diggers. Several miners report that these people trespass and steal from the mining operations at night, and many are confrontational. If you come across any of these people, you are asked to report them to the mine owners and the local authorities.

Another thought comes to mind if you are collecting someplace other than the fee pay mines. During certain times of the year, there are hunters in the woods. Squirrel, deer, turkey and other game seasons bring out great numbers of excited people who hunt with bows and arrows and guns instead of rock hammers. Bright orange vests are required for hunters in the woods during certain of these seasons. Deer hunting is such a big event in Arkansas that many of the rural schools let out for the first day of deer season.

If you stop by WalMart in Hot Springs, or other sporting goods stores, you can get a card that lists the dates of the hunting seasons, and you might want to pick up an orange vest or cap if your time in the woods overlaps with the hunters. I'm reminded of one of those timeless truths circulating on the internet, "Why do they call it tourist season if we can't shoot at them?"

The Arkansas Game and Fish Commission can also give you season dates. Check their website for current information http://www.agfc.com/hunting/

Digging your own quartz crystals can be as simple as walking over the ground looking for sparkles to using heavy equipment.

3. Collecting the Crystals

KNOWING WHAT to look for is half of the success in finding a crystal. A good eye, knowledge of the area, and the right tools are also essential. Luck and perseverance never hurt, either.

We are often asked, what crystals are worth picking up? A good philosophy is, if you pick up the crystal and like it, it's worth bringing home.

First time collectors usually go for quantity instead of quality, and after they get home and clean all their finds, they may realize some of the pieces were not as good as they hoped. First of all, when you go into a rock shop and see all those beautiful sparkly clean crystals on display, you are going to think you will be able to go the mine and pick up the same things. Well, you might be able to, but they will not be beautiful, sparkly, nor clean. Crystals in the ground are muddy, often brown stained, and camouflaged.

It is very good to go into the retail shop of a mine to get an idea of the quality material they produce, but realize that you are looking at months, and perhaps years of mining to get the quality and amount of material it takes to stock a store. The owners have also put in many hours cleaning, trimming and preparing these pieces.

The easiest way to find crystals is to walk over the ground and look for a shiny reflection. Can you also find the small crystal point in the upper right of this photo?

Expectations

It is easier to find lesser quality pieces than high quality specimens, but don't let that discourage you. If you've ever gone fishing, you know why they call it fishing instead of catching. The big ones sometimes elude us, but the more we know about what we're trying to do, the better the chances of a successful trip.

Your past collecting experience will affect the kind of day you have at the mine. One digger I talked to said they were getting very large exceptional pieces on a certain day, when a novice came up, picked up some of those kinds of pieces, then threw them back on the ground. His thinking was that because they were everywhere, they must not be very good, and he'd look for something better.

Another lady threw back all her small crystals, believing only the big crystals were good. She wished later she had saved them. Large and small crystals both can be nice.

Here's another thought on bringing home the smaller crystals. Even if you don't think they are very special,

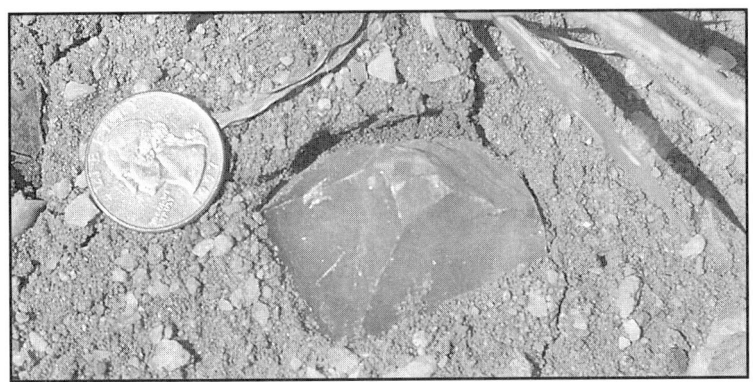

Above: If there was a recent hard rainfall, you may be lucky enough to walk over a tailings pile and see the tops of the crystals sticking out of the ground and easily pick them up. A hard rain washes smaller crystals almost completely out of loose dirt.

Below: Working the tailings pile of Miller Mountain mine. Garden tools do a good job of moving dirt and exposing the single crystals and clusters. The material is brought up out of the mine pit and dumped for diggers to work through.

If the sun is not at the right angle to reflect a gleam, a crystal in the ground can also appear dark. The dark area in the center of the photograph on the right was the only part of a crystal exposed in a tailings pile. Laurie carefully dug the crystal out. After the crystal had most of the clay wiped off, below, Laurie found she had a nice hand-sized single crystal. When the crystal was rinsed, it was very clear with an undamaged point. There is a note of caution in the text about rinsing crystals and putting them in the sun to dry. From the Crystal Hill mine.

chances are you have a neighborhood kid who would. To this day, at my mom's home in one of her sewing machine drawers, are the very first crystals I ever got. Old Mr. Pittman gave them to me. There is a lot I don't remember about being seven or eight, but I never forgot about my crystals, and I probably wouldn't have even remembered my grandmother's friend except for his generosity!

When taking kids with you on your collecting trip, show them right away what you are looking for. Otherwise, you will be bringing home a truck full of yard rocks.

Show the youngsters these are what you look for.

Advanced pebble puppies

There are a few technical terms advanced pebble puppies need to add to their vocabulary. One is Trashite, another Leaverite, and another High Grade. The first term, trashite, is a mineral specimen of no value. Trashite should always be left at the collecting site (in situ, as geologists say). That way it will never interfere with your collection. Leaverite, used as a noun or a verb, means "leave it right where you found it". It is also a specimen you don't need to waste your time on, but let someone else take it home to their waste pile. High grade is a verb, meaning pick over your collection with a critical eye and select the best pieces, while giving away, selling, swapping, etc. your lesser pieces. High grading is a good practice at the collecting site, too. It prevents you from having piles of Trashite and Leaverite around your yard. As your collection grows, aim for quality, instead of quantity. Your significant others will appreciate it and you will feel better about your collection.

Surface collecting

Small crystals laying on top of the ground are worth picking up if they are clear and not broken. If the point is badly damaged or missing, that's leaverite, so don't bother with it. Iron stained and muddy crystals are worth picking up if, again, they are not badly damaged, because they can be cleaned as described in the next chapter. Quartz is highly resistant to weathering, but no match at all for a bulldozer or trackhoe. As the size of the crystals increase, you have to make a decision for yourself as to how much damage you can tolerate on the points. We see many, many single points that are broken from mining, and most of these we leave at the site. Some break our hearts because they would have been such fine pieces if the point hadn't been so badly dinged. Depending on what you plan to do with your crystals, you'll need a discerning eye to pick out the ones to take home with you.

If you want to make jewelry, you'll be looking for something much different than if you want a landscape piece for your rock garden. For persons who use crystals in healing work, smaller crystals will have the same features and characteristics as the larger ones, and it's easier to find small ones just because there *are* more small crystals than big ones.

We've found some large, water-clear broken pieces of crystal that were hard to leave behind, and if you come upon such a piece yourself and can't part with it, consider taking it and having it made into a sphere or cut stone. There are lapidary people who have special machines to do such a thing. If you visit the tourist shops in Hot Springs, you might see "Hot Springs Diamonds" which is a local name for stones cut from quartz. When they are faceted like a brilliant cut diamond they can be attractive, but lack the fire and sparkle of a real diamond.

When surface collecting or going through tailings, you'll have the best success by either covering a lot of ground or moving a lot of dirt. My favorite technique is to walk over the ground in different directions looking for surface crystals. When I see one, I use my toe to check for buried parts in the ground, and if need be, I pull out my garden tool and then bend over to dig it up. Freshly turned dirt is a good place to look, and recently turned dirt that has been rained on is the best. Another good place to look for crystals is in the tire tracks where the trucks go. Weathered areas are easier to look over.

The Elderhostel classes we teach in Hot Springs always make a field trip to one of the local mines, and it's amusing to hear all the grunts and groans coming from the participants. It's hard to say if the knees or backs go out first, but there certainly is strain in repeated bending to get many crystals. I heard a description likening all the group bending and grunting to some kind of line dance, and if accompanied by good music, could be very entertaining!

What I'm getting at is that if you have a physical disability, you might want to bring a sit-upon of some kind, lawn chair or bucket perhaps, and park yourself in one place to work with your digging tool. A few mines even have handicapped accessible areas, but they are not called that. One day I saw two elderly people on scooters working the ground at the Ron Coleman mine. That's not possible everywhere, but an alternative for those with disabilities.

Working the Veins

Finding a pocket

Searching for pockets takes a keen eye. White milky quartz veins are easy to see and may lead to a pocket. Keep

Mike found a vein of dark orange clay. He spotted the tip of a crystal sticking out. Using his pocket knife, he worked the dirt loose around the crystals and popped them out.

With just a couple of minutes work, his time paid off with a handful of crystals that will clean up nicely. Some veins and pockets yield very large crystals. From the Starfire mine.

a look out for them.

Sometimes when you are working on a vein you may open up a pocket or vug (a cavity into which all the crystals point to the middle). Often such pockets will yield very good crystal specimens, but you must take considerable care in working the crystal. Quartz is brittle and nothing sounds worse than trying to recover good specimens and hearing the crystals grinding or crunching against each other! I'd rather leave them in the ground than break them all to pieces trying to get them out.

When a pocket is discovered that looks good, dig down beside it and come in from the side to collect the material, if possible. Use the prying tools to wedge into a crack and

The mine owners expose pockets using heavy machinery, then begin hand work with a pry bar, such as you see in this picture. Tim Hill, owner of the Crystal Hill mine where this photo was taken, has stern words for anyone who carelessly thrusts a steel pry bar into a pocket, as doing so can damage crystal and break off the points. In a miners words, that's the "sound of the price of quartz going down."

ease the sections out. You'll have a lot less damage than if you stick your bar directly into the cavity. Most of the pockets are tightly packed with clay. Don't try to remove the clay right then! It's Mother Nature's packing material. You can clean the crystal later. Wrap the crystals or the cluster with a single sheet or half sheet of newspaper and pack in your box or bucket to take home.

Hot weather precautions

I have had some interesting conversations with various collectors and learned some additional facts. Ken Silvy, who digs herkimer diamonds (double terminated quartz crystals) several months a year in the Middletown, New York area, told me about how he has to handle the crystals when he takes them out of the ground. He keeps them in the cold clay and places them in a bucket of cool water in the shade. He says he has seen many first-time collectors ruin crystals by wiping off the clay and putting them in the sun to dry. POP! CRACKLE!! The crystals heat up rapidly on the outside while the inside is still cold. Thermal stress builds up until they fracture. They must warm up very slowly to prevent the stress building to the point of overcoming the internal strength of the crystal.

Some time later, Ron Coleman told me an interesting and somewhat sad story relating to a specimen from a tremendously large pocket of crystals he and his brother Jim dug near Jessieville in Garland County, Arkansas. A few of these crystals were 3 feet in length and 1 foot in diameter. One in particular was clear from the termination to over one-half its length! Within a minute after it was removed from the cold ground, a fracture appeared about a foot from the base. They quickly covered it with blankets, but it was too late. While they watched in amazement (and disappointment as they saw $$$'s flying out the window) the fracture grew up the length of the crystal to within 6 inches of the termination. To prove his point, Ron took me over to Jimmy's backyard, pulled back some blankets and showed me the specimen!

Safety precautions

When you get to the area you are going to collect, you'll want to be sure that all the people in your group, especially the youngsters, understand the basic safety rules.

Stay away from walls where rocks can fall on you, don't get close to edges of pits that you can fall in, and stay away from edges of mountains you can fall off of. Wear your safety gear when you need to, and don't mess with the mining equipment the operators might have there.

Chances are someone in the group will get cuts or scraped knuckles, and maybe a mashed finger. Your water bottle and first aid kit will be able to clean and handle these injuries, so be sure and have them with you.

An ounce of "safety-first" beats a ton of "I-wish-I-hadn't-done-that" later!

This is not a trick question. Why is it a good idea to wear a hardhat in a mine? Answer: to protect your head and face from falling rocks. In the business, all that big highwall overhead is called "headache rock."

4. Cleaning and Trimming

NOW THAT you have collected some nice specimens, and because they are still dirty, you need to know how to clean them. Don't do this in your sink because the clay will stop up the drain. If you have just a few pieces, use an old toothbrush to get the clay off. Otherwise, start by building a 2 by 2 foot framed 0.25 inch mesh screen. You can get this type of screen, called hardware cloth, at your local hardware or farm and garden store. Remove the newspaper wrapping and let the specimens dry on the screen for a couple of days. Keep everything in the

Screen table and basket of mine run quartz ready to wash.

Repeating the rinse and dry cycle will get the bulk of the red clay off the crystals. The size of the mesh needs to be small enough that the small crystals don't get washed away.

shade to prevent the crystal from heating up too rapidly in direct sunlight. When the clay is well cracked and dried, rinse with a garden hose. Let dry a couple of days and repeat the cycle.

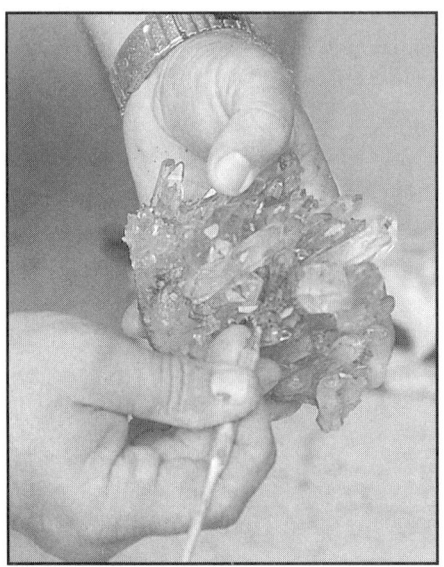

A dental pick is ideal for getting clay out from between the points of a cluster. The more of the clay you can remove before you start any acid treatment, the more efficient the time in the acid bath will be.

Toothpicks, bamboo skewers and other small implements make great clay removal tools.

Some key points
There are several key points to preparing quartz to be cleaned. First, you need to remove the clay. This is accomplished by cycling the specimens through several wet and dry periods to loosen and wash the clay away. You may have to use a pressure washer to remove the last of the clay. Remember: this first step is critical.

You do not want to have to clean the material several times but that is exactly what you will have to do if the clay is not completely removed before the first acid cleaning. Trim your specimens to the size and shape you want *before* cleaning in acid. Remove the dinged or broken portions to save the time of re-cleaning after trimming.

Trimming

Getting rid of extra matrix or removing unattractive parts of the specimen will enhance your pieces, sometimes turning an ordinary piece into a real show piece. It takes study and courage, but the rewards are worth it.

Ever wonder what you are going to do with all that leaverite and trashite you hauled home and haven't gotten rid of yet? You can use some of it to learn about trimming specimens. After reading this chapter, try trimming out a pocket of broken crystals, just to get the experience!

Trimming usually begins in the field where you find the specimens. I usually carry a 8-pound sledge hammer and a 4-pound crack hammer with me wherever I roam. (See also Ch. 2 about collecting tools) These are generally sufficient to get the specimen down to a size I can bring in from the field. Also, many materials will not hold up to this type of trimming (the shock from the hammer blows break the crystals loose from the matrix) so it becomes evident that only through experience will you learn what can withstand collecting and what to leave for someone else to waste their

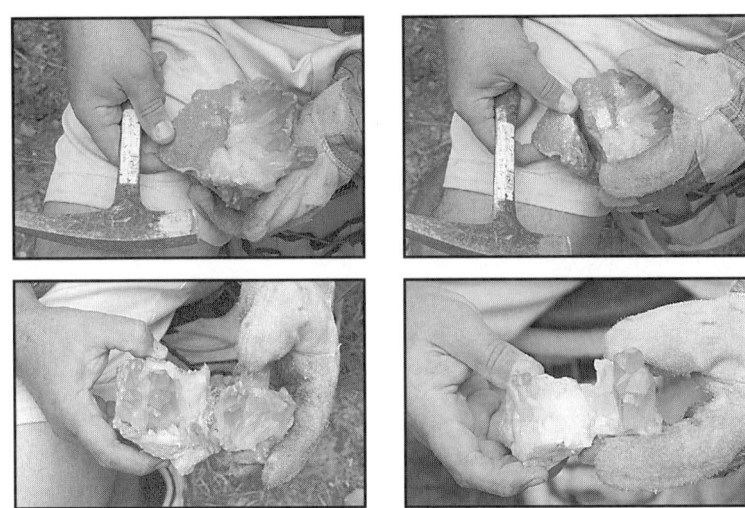

A double hand size piece is trimmed with a hammer. The first blow removes only matrix, and successive whacks get closer to the desired part to save. This fairly large piece had only a small area of good crystal, which was turned into an attractive specimen with careful trimming. A leather glove protects the hand holding the specimen.

Trimming tools: A screw-type trimmer, tile nippers, ball peen hammer and section of railroad rail. Pinching-type tools give better control than breaking rock with a hammer.

time and energy on. A knowledge of the matrix and its peculiar nature is essential. You will ruin some specimens learning how to trim, but hopefully, you will also learn how to greatly improve some of your specimens.

Trimming tools to use

There are many tools that can be used to trim specimens. You are only limited by your pocketbook. Simple hammers, chisels, and tile trimmers may be supplemented by screw-type pressure trimmers, hydraulic pressure trimmers, and/or diamond saws. (Note: a peculiar trait of veteran mineral collectors is that they really dislike a saw cut surface on a specimen. If you must use a saw for trimming, be prepared to take some flack from some collectors.) Never trim a specimen just to make it set up nicely. You can always get a stand to accomplish that goal.

Go easy. Shock is the great enemy of any specimen. Anything that can be done to reduce shock will give you better odds of accomplishing your task. I once cradled a cabinet-sized specimen in my lap and held it with leather gloves as tightly as possible while a companion trimmed over half the matrix off with a 4-pound sledge hammer. The weight went from about 15 pounds to about 8, while the wholesale value went from $225 to $650! It was hard work and took about 10 minutes of concentrated effort, but was a most interesting experience. The collector, an expert in trimming this material, had saved the piece for almost a year because he knew I would be trusting enough to help him, and because he could not do this by himself.

The goal

The purpose of trimming is to remove excess material to improve the overall quality and to remove damaged areas detracting from the overall aesthetics of the piece, thus improving the value of the specimen. If you cannot accomplish these goals, then do not trim the piece.

Matrix vs. crystals

Matrix has its own properties, independent of the minerals you desire to collect. It may be soft and punky, brittle and highly fractured, compact and hard as the hinges of Hades, uniform and predictable, or any combination of these characteristics. Keep in mind the strength of the crystals you are trying to recover. Are they firmly attached or about ready to fall off? You can quickly learn these properties (of both matrix and crystals) from a poor quality specimen by tapping on it with a regular rock hammer.

I can make a few recommendations for trimming the base rock from the crystals. Then it's up to you to try, and through your efforts, gain the necessary experience.

A larger specimen, below, had some damaged crystals. It could be made more valuable as higher quality smaller pieces by careful trimming. This piece was trimmed with a rock hammer while it was carefully cradled in the lap.

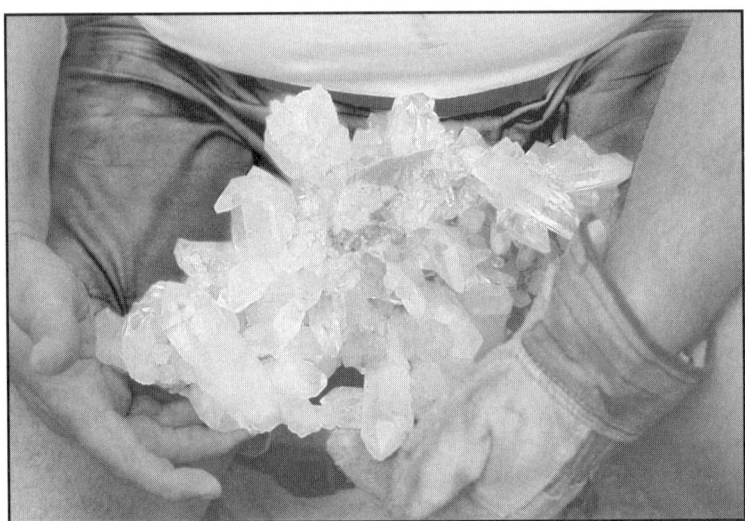

Two kinds of matrix

There are two types of sandstone matrix, although you may find that one will grade into the other. The first is hard and uniform, often with pre-existing fractures, and the second is soft, granular, and tends to come off by rubbing with your fingers. Each of these types of matrix have their own characteristics.

When I work with hard matrix, there is little control as to the size and shape of clusters as they come out of the ground. I have to work away from the pocket somewhat or all I will accomplish is breaking off individual crystals or smaller clusters. Once I have a specimen with hard matrix, I closely examine it to look for pre-existing fractures. If there are any, one hit from a hammer will generally break the specimen along them. This may be good or bad, depending what you will have left. If the matrix is solid, any trimming must be done gently and not at right angle to the vein. Hitting in that direction will pop crystals off the specimen due to the shock of the blow. If I want to break the specimen in a given place, such as to remove a section that is damaged, I would cradle the specimen in my lap to

Results of trimming the large specimen on the left page: two quality specimens and about 50% waste (not shown).

absorb the shock, and hit on the already damaged spot. Now keep in mind this will not work if the damage is in the middle of the specimen, only if it is near the edge! I wear jeans when I do this and place something protective between my clothing and the specimen, like an old piece of cardboard or a carpet remnant. Trimming in this manner with a hammer takes experience to master.

Soft sandstone matrix is a completely different situation. If the matrix is friable (loose like sugar) and the sand can be removed by abrasion, then I might use a grinding wheel or even an old pocket knife to scrape off the loose sand. Once the specimen has been cleaned, the remaining binder of the sandstone may be gone and the loose sand easily removed from the back of the specimen. Such specimens can be easily backlit for a dramatic effect, but they are particularly fragile. I have purchased several pieces which had a lot of loose matrix, brought them home and finished the matrix removal with a dental pick or electric vibrating needle. One such specimen looks like an inside out geode with the hole where I removed the sandstone matrix big enough to put my hand in! This type of specimen is easy to recognize for it is always shedding its sandy matrix .

Pressure trimmer a boon

On some specimens of either type of matrix, a pressure trimmer is essential. The effectiveness of this type of trimmer comes from the highly directional planar pressure (pinch) it generates in the matrix. See photo on page 48.

Several brands of small screw-type trimmers are available. The size of specimen you can trim is limited by both the distance between the vertical rods, the length of the rods, and the strength of the matrix you want to break. If too tough, you could strip the central screw's threading. So, you have to use common sense or you will ruin the machine. These trimmers come in small and large versions.

Know when to quit

The final trick of trimming comes from knowing that you have done everything you possibly can to improve the specimen and that it is time to quit. I have witnessed some extremely poor examples of trimming and some spectacularly successful ones. One example concerns a wavellite specimen purchased from a local dealer at Mount Ida by a university professor. He agreed to let a friend, known for his trimming expertise, remove some matrix right at the guy's shop. One whack and the specimen was only one third as large and probably worth 5 times what the prof paid for it.

In another example, a friend loaned me a spectacular variscite specimen for a display. I noticed that a small knob on the backside of the specimen appeared to be loosely attached. One tap on that knob to remove it and the specimen split in half! Fortunately for me, most of the mineralized portion was on one half, having not split down the middle of the cavity. The lesson? No matter how much experience you have, never try trimming someone else's specimen without their permission!

Although hard to see in a black and white photo, the center cluster is noticeably darker because of the orange iron-oxide staining on it. This stain is best removed using an acid bath. A well-trimmed and cleaned cluster is worth much more than a rough piece of quartz.

Back to cleaning

Once you have the specimens prepared for acid treatment, you must consider the situation. Do you have small specimens and just a gallon or two of crystals, or do you have some big pieces? Maybe you got lucky and have one piece that would fill a 5-gallon plastic bucket!

Removing the iron oxide

Your pieces may have a light iron-oxide staining or a heavy iron-oxide coating. The light iron-oxide staining will be translucent to light, while a heavy iron coating forms an opaque crust.

The next step in cleaning is getting rid of the rust. If the crystal only has a very light iron staining, then a few days soaking in a weak oxalic acid solution in a covered plastic bucket will remove the orange-brown color. If the iron staining is heavy, then you must cook the quartz in an acid solution. People have many ways to clean quartz, all involving basically the same scenario. Your specimens may be coated by iron or manganese oxides, with or without clay, and you should remove the staining so the specimens are as clean as possible.

To clean crystals without heating them, let them soak for 5 or 6 days in the acid bath. Most people who are serious about quartz will use the hot method that takes 3 days. Many contraptions have been used to cook quartz, from fish fryers to custom-made acid baths with hoists over the vats.

Thermal shock

Heating and cooling too quickly will cause fractures in the crystals. To prevent cracks, remember these points in handling your pieces. Start the cleaning in cool water, and after cooking, turn off the heat and let the crystals cool to air temperature before you take them out of the cooker. Putting a cool crystal in a hot pot will also make it crackle. In the same kind of situation, resist the temptation to take a piece out of the hot acid just to see how it is doing, for the sudden cooling will make it crack. Just one experience of this happening gives a lesson you'll never forget!

Tips for using oxalic acid

The most commonly used, readily accessible chemical for cleaning quartz is oxalic acid, which may be bought as a white crystalline powder. It may be purchased from many mineral dealers in Arkansas, especially those who specialize in quartz. Buying the acid from a quartz dealer is a good idea, because if you go to a chemical supply company you will have to buy an enormous quantity of it.

Don't use expensive reagent or chemical grade oxalic from the pharmacy. Many quartz dealers provide written instructions and sell it by the pound or package for around $3 to $4 per pound. How much do you need to buy? It depends. Did you get a 5 gallon bucket of quartz specimens or a 3/4 ton truck bed full? A five gallon bucket of crystals might take 1 to 1.5 pounds of acid to the water if the crystal is really dirty. And by the way, when you purchase oxalic acid for cleaning your crystals, get a receipt that plainly states what this stuff is and who you bought it from, because it looks a lot like drugs to police!

When mixed with water at a measure of a few ounces per gallon and then heated to just below boiling, oxalic acid is capable of removing all but the most stubborn iron staining. It is a weak organic acid, but don't kid yourself, it will hurt you—especially if you breath the fumes. So use it only outside in a protected and well vented area, where no children may gain access.

A good recipe to begin with is one pound oxalic acid to 2 1/2 gallons of water. Start with a weaker solution first and build its strength if needed. Also, the acid solution will not be used up until it turns dark emerald green by becoming

No need for expensive equipment to clean your crystals. Yard sale crock pots are perfect for cooking single crystals and small clusters. Please note that once you use acid in the cooker, it is no longer safe for food or fit for use in the kitchen.

saturated with the removed iron, so you can reuse it by adding a very small amount of fresh water and powdered acid to the old solution. Remember: it is an acid, though a relatively weak one. Do not leave this stuff where kids or animals might get into it. I usually wear dish washing gloves when working around it. Even though it will not burn skin, it will let you know if you have a scratch or cut. Wear your safety glasses and take care not to get any acid in your eyes. I always keep a garden hose handy so when I get a splash, I can rinse it off immediately.

Experiences with cookers

To clean small pieces, you need to search for cookers at garage sales. Whenever you find a crock pot (the slow cooker ceramic-lined type) for $4 or less, buy it! Dedicate it to cleaning crystals, because its kitchen days are over once you've used acid in the cooker. You can get some 10 to 12 processing cycles before the acid finds its way through hairline cracks in the ceramic inner glaze and corrodes the heating element. But that's okay, if you got 10 gallons of small quartz specimens cleaned, then it's worth it.

Place the specimens in the crock pot, add cold water, then a couple of ounces of dry oxalic acid and top off with cool water. Be sure the water is above the crystals, because any crystal sticking out will not get cleaned. Cover with the glass or plastic lid, plug in and set the temperature control to low. Check this every two days and add a little warm water as needed to keep the crystals submerged. DO NOT DO THIS IN THE HOUSE. ACID VAPORS ARE POISONOUS.

After about a week, turn the crock pot off and let it cool down overnight. Do not get too anxious to pull the crystals out while they are hot or they will shatter from the thermal shock. Then remove the specimens and rinse them thoroughly. If your specimens begin to grow a white powder as they dry, place them back in a clean crock pot, add water and 1/3 cup of powdered agricultural limestone or lime, and cook overnight. This will neutralize the remaining acid as it comes out of the pore and cracks of the specimens. If this does not work to get rid of the white powder problem, then you will need to cook them a second time in clean water with ag-lime as a neutralizer.

Disposing of acid

To dispose of a volume of spent oxalic acid (it will be a dark emerald green color from the dissolved iron it contains), add agricultural lime (CaO) that you can purchase at the garden supply stores to the liquid until you get no reaction. Then it will be neutralized due to the formation of harmless calcium oxalate. You can dump it on the ground where you wash your crystals with a garden hose. That way, the next time you wash rocks, or it rains, the material is diluted. If you lived in town or in an apartment, just take a funnel, pour it in a 1-gallon milk jug and put it in the trash or in a dumpster. Since its neutralized, it is not considered a hazardous material, and since it is water-based, it is not flammable.

Cooking big pieces

Now, how about your big specimens or your 5-gallon bucket of hand-sized pieces that would take too long to clean in the crock pot 2 or 3 at a time? There are many ways to cook big pieces. By years of experimentation with different types of cookers, I think I have found some methods that work satisfactorily. You will need two things that can be an assorted variety: a container for the crystals and a heat source. We've seen wood fires, propane burners, and electric heating elements used.

I got the idea for my first big cooker by observing Sonny Stanley's cleaning operation. He uses a wood fire.

I had a tank made at a metal shop from sheet steel. It had 18-inch tall legs and the tank measured 3 X 3 feet and 2.5 feet deep. Two expanded metal screens fit inside this cooker, one stacked on the other for 2 layers of quartz. It held about 10 gallons of crystal at a time and once hot it stayed hot for about two days with no fire underneath during the spring, summer, and fall. I had a cover for the top to prevent too much evaporation loss, and the entire cooker set on a small concrete slab for a level base.

I fired this cooker with oak firewood overnight. It worked very well for 2 years, probably 15 cooking cycles, but then a weld seam sprung a slow leak. I had it rewelded, but knew that the acid was slowly eating the tank up from the inside out, so I sold it. It cost around $125 to have made and I sold it for $75. At 150 gallons of cleaned crystal for $75 final container cost (the wood was free), it was pretty economical to operate. It did require some periodic tending several times a day.

A second kind of cooker is made from a steel 55 gallon drum, like you would use for a burn barrel. Have someone cut the thing off 18 to 24 inches high. Place this on a brick or steel stand. Old kiln-liner brick can withstand very high temperatures and are perfect if you can get some. Or you could have a welding shop make a free-standing base with legs about 15 inches in length. Again you need a top to prevent rapid evaporation. You can fire with wood or a propane fish-fryer kind of burner. My unit worked well for a couple of years. The barrel only cost $10, but it finally rotted out from the acid.

Another type of homemade cooker may be the best. It is simple and is built on a double boiler principle. The crystal is placed in a 5-gallon plastic bucket with water and acid like usual. Then the bucket is set on bricks in a larger metal container. The lid of the plastic bucket is set on but not snapped down, as the pressure will make it pop off! Add water to the large metal container, and cook with either wood or propane, keeping the water just simmering. You must add water as it evaporates.

One person I know cooks a lot of crystal this way and he uses an old bathtub set on concrete blocks. He plumbed the drain with a pipe which extends out the side of the unit and ends with a gate valve. Three to four 5-gallon buckets at a time are cooked with wood in this setup, for up to 10 days. He sets his buckets on bricks to prevent them from melting. Since no acid is in the tub, it has lasted for years.

A yellow crust problem

If you use oxalic acid, sooner or later you will have a batch that gets a yellow crust coating it. I do not know exactly what causes this problem, but I have an idea and I will tell you how to remedy it if it does happen. I encountered this problem most often when I used a white plastic bucket, almost never with a colored plastic bucket (like red or black). I think the sunlight causes some type of reaction which allows this yellow iron compound to precipitate out of the acid solution. When this happens you will think "all my crystals are ruined!" Don't get too upset. Remove the crystal and place in another bucket. Fill near the top with water and add some muriatic acid. Be careful to always add acid to water, not water to acid! Let set over night and the yellow staining should be gone. About a pint of muriatic to a five gallon bucket of crystals should be enough. If the yellow stain is not gone, then add another pint and wait another day. Be very careful with this acid. Muriatic is a commercial grade of hydrochloric acid and is available at most hardware stores. When working with it, keep a garden hose nearby and turned on at the faucet for safety. Wear safety glasses and rubber gloves when pouring, and only work in the open air. The fumes from the full strength acid in the bottle will damage your lungs!

Ultrasonic and needle spray gun cleaning

If you want to get really high tech for cleaning some quartz, you may want to consider two separate pieces of equipment: an ultrasonic cleaner and a high-pressure needle spray gun. I would not plan on using this equipment for bulk cleaning of specimens, but for individual pieces, it may come in handy.

An ultrasonic cleaner works with high frequency sound waves. Cavitation occurs when millions of tiny bubbles form and then burst. With ultrasonic cleaning, the water-based solution needs to be able to wet the sample (wetting agent) and to lift the material off with a surfactant. Clay may be removed from the tiniest cavities by cleaning in an ultrasonic tank. These machines come in various sizes, most machines with a submersion tank larger than a quart cost over $100. Ultrasonic cleaning solutions can be purchased as liquid concentrates that you simply mix with water and use. There are dozens of well-known ultrasonic manufacturers and retailers. Either do a search on the Internet for ultrasonic cleaners or check the ads in national mineral and rockhounding magazines. Be sure you get a wire basket for the tank, also a tank cover is handy.

A high-pressure needle gun is really useful for cleaning all types of individual mineral specimens, not only quartz. But the cost is prohibitive to the average collector—some $350 to $400! This is a machine that operates like a car-wash sprayer, but it is hand held and puts out a needle-thin stream at around 160 to 200 pounds per square inch. You must be careful using it or you will shoot a hole in your finger! A needle sprayer is very handy for cleaning the tiny crevices of mineral specimens, particularly the ones that you can not quite get clean with a needle probe. Wear safety glasses or some type of eye protection because when you blast that clay out of the specimen it flies everywhere, including all over your face. I have come in from cleaning specimens with one of these sprayers and barely been able to see out of my prescription glasses due to of all the mud that splashed back in my face!

Pressure washers and ultrasonic cleaners get the crystals really clean.

5. Places to Collect

JUST A FEW years ago, there were more publicly accessible places to collect rocks than now at the turn of the century. Some areas have been cleaned out, others have been reclaimed, and still others have been put off limits by the owners.

On the whole, Arkansas is a friendly place, and if you haven't been here before, you'll notice that people sitting on their front porch will wave to you, as do people you pass on a dirt road. There are locations, though, where rockhounds have become a nuisance. Some local collectors tell stories of landowners wielding a shotgun and running off trespassers. With these thoughts in mind, before we tell you where to collect, we'll tell you the places not to collect. If you see trees or fence posts painted with lavender or purple paint, that means the land has been marked as No Trespassing. You may even come across particular landowners who think they own the state highway right-of-way. If these folks show up, the better part of valor would be for you to leave.

If the land is properly posted with purple paint, no warning is necessary by the landowner. Arkansas law carries a $500 fine per person on the first offense. We'd rather you enjoy your time here and not meet our law officers.

Places not to collect
- National and State Parks
- Lake Ouachita islands and shores (Corps of Engineers)
- Private property without permission
- Contract mining areas and staked mining claims
- Areas marked with purple paint on the trees or fence poles

PLEASE BE AWARE that knowing about crystals and where they can be found does not grant the right to trespass on private property or mining claims. Even if you consider taking some samples as "just collecting", property owners might consider your actions trespassing and theft. The status of mineral collecting on National Forest and Corps of Engineers land is changing, and access can be very restricted. Please check with the appropriate landowner, lease or claim holder, or district supervisor before attempting to enter a collecting locality.

Free Places to Collect Crystal

Ouachita National Forest Land

The forest's complex geology offers unique opportunities for both professionals and rockhounds. Rockhounds are welcome to pick up quartz crystal off the ground for personal use but not for sale. If you would like to use big tools to dig for quartz, such as for a rock club field trip, or mining crystals for sale, contact the local district ranger.

A good person to contact is the US Forest Service Geologist:

John C. Nichols
Ouachita National Forest
P.O.Box 1270, Hot Springs, Arkansas 71902
phone: (501) 321-5285 FAX: (501) 321-5353
website www.fs.fed.us/oonf/minerals

The National Forest Rockhounding Policy:
• Rockhounding may take place anywhere in the forest *except* in closed areas (i.e. wilderness) and from quartz contracts unless permission is obtained from the contract holder.
• Crystals are for personal use only and not for resale.
• Only hand tools may be used.

John Nichols, the Forest Geologist for the Ouachita National Forest, states in his paper on Rockhounding the National Forest Land, the key to mineral collecting and rockhounding on our National Forests is communication: making contact and talking with the local District Ranger. Not only will the collector find out what the policies and rules are, but they will discover that the Forest Service Personnel can provide many other practical resources such as maps, information on camping and other recreational activities, and other items of interest on the Forest. Some Forests have Geologists or Mining Engineers on their staff, usually located in the Forest Supervisors Office. Others rely on the expertise of Geologists in their Regional Offices. The Forest or Regional Geologist can often provide information on Forest policy and procedure, facilitate communications with the District Ranger, and provide invaluable mineral, geological, and reference information to the rockhounder.

There are two free areas in the Ouachita National Forest open to rockhounds:

Crystal Vista, Montgomery County
Womble Ranger District, PO Box 255 Highway 270 E
Mount Ida, AR 71957 Phone: (870) 867-2101

From Mount Ida, travel south on Highway 27 for 3.8 miles and turn left (east) on county road 2237 (Owley Road). Continue on 2237 for 4.1 Miles to Crystal Vista trailhead and parking area. Fees: none; Facilities: none; Open year round.

Crystal Vista is a 4-acre clearing in the trees on top of Gardner Mountain. Access is by a steep trail, about one mile long, and takes about 25 minutes to hike.

Once a commercial crystal mine, quartz crystals can easily be collected off the surface of the ground. A short walk to the top of the ridge has a spectacular view of Lake Ouachita and the surrounding landscape. This land has been reclaimed and is closed to vehicle traffic.

Crystal Mountain, near Lake Winona, Saline County
Jessieville Ranger District, Jessieville, Arkansas
Phone: (501) 984-5313.

From either Hot Springs or Benton, take Hwy 5 to Hwy 9. Turn north and go to Paron. At Paron check your mileage and continue north 4.5 miles to Forest road 132. Turn west on 132 and drive approximately 10.5 miles to Crystal Mountain. An unmarked access road to the top of the mountain spurs off to the south. Take this spur about 1/4 mile to the top. The last part of the road is rough.

Seemingly way, way out from anywhere, this location has a view so breathtaking it is worth the drive just to look around. We took our cub scouts to this location and found a good number of small single points that could be scratched out of the surface. We also found a lot of trash that we picked up, testimony to careless rockhounds who have no regards to their surroundings. We've written elsewhere, and say it again, pack out all your trash when you go collecting!

Fee Pay Mines

At most of the fee-pay mines, the mine operators collect from veins and pockets in the walls and floor of the mine, usually a pit. The mine operators don't get ALL the good quartz! They load up the tailings and clay from the pockets and bring the material out and dump it in the fee area.

With some effort to dig through the tailings, you can collect nice material. Quality and quantity of specimens vary greatly from mine to mine and even at each mine on a daily basis, depending on what the mine operators are digging in. If they have a lean collecting day, so do you. If they have pockets galore, then you'll get some goodies also. We can't recommend any particular mine over another, but several mines do allow you to dig on the veins. When you get to do this, it can be hard work, but very rewarding with good specimen material. Or you may draw a blank day, just like the miners.

I have an aunt who says the reason the south lost the war is because the roads are so poorly marked. All of the mines are on dirt roads in the middle of nowhere, and you are going to feel you are really in the backwoods when you visit some of these places. Only a few mines have any facilities nearby, so you are well advised to take snacks and drinks with you. In many of these places, there aren't even conve-

Mines in the Ouachita National Forest are clearings in the woods, usually a few acres. The scenery and drives are as much of the adventure as collecting the crystal. View of Fiddler's Ridge mine, looking south from Fisher Mountain.

nience stores close by on the main roads, which is a good reason to have your car's gas tank full and do like the Boy Scouts—Be Prepared. The Forest Service roads are dirt and well maintained. Occasionally though, we've found signs missing. If you've got a map, gas and groceries, the worst that would happen is you'll have a long scenic drive that would cut into your collecting time. Although the Forest Service roads are in good shape, some of the access roads to the mining areas themselves are rough. We visited most of the mines in our Honda Civic and made it fine. Check with the mine owner about the entrance road.

Pay before you dig

Crystal is a synonym for money to the mine owners. Most, but not all, of the mines have a retail shop. Not all the retail shops are close to the mines; Stanley's, for instance, is pretty far away. You will need to pay your entrance fee before going to the mine to dig. Generally you buy your permit at the shop, then go to the mine to dig. Some owners make different arrangements, such as meeting you at the mine or scheduling with the mine foreman for your trip. In the places where you are allowed to dig in the pits or on the veins, you will be asked to sign a liability release in case you are injured. Since there are various arrangements at different mines, especially in the Mount Ida area, contact the individual mine owner to get updates. Spend a little time at the miner's shop looking at the material on their tables. You can then get an idea of what you can find, and you will also know if what you got was ordinary or good.

A large number of the mines are family run operations. Some show prosperity, and all show that mining is hard work. The nature of the crystal business includes wholesale as well as retail sales, and as these people supply rock shops around the country, many dealers travel to regional shows at certain times during the year. The smaller mines will close to go to these shows, such as the international gem and mineral show at Tucson, AZ in early February, the Denver show in early September, and other weekend shows at different times and places. You may well be able to see these people at the major rock shows around the nation. If you have difficulty getting in touch with any of the shops, the show schedule may be the reason. The operators are business people and they have decisions to make. It might be that working the mine at a certain time may not be a priority for them.

This information was current as of the Fall, 1999.

Crystal Heaven, Montgomery County
Contact: Stuart Zove
Address: 104 Sunrise Hills Drive, Mount Ida, AR 71957
Phone: (870) 867-4625
Website: www.crystalheaven.com
Fees: $20 per adult, with a $5 discount per person for groups of 8 or more. Also, they will close the mine to other collectors for the day with a group this size. Kids 14 and under are $10 per day. You may dig in the pit with the miners. They say they will even use their equipment to expose some crystal for you.

The word is that about 50% of the crystal produced from this mine contains phantoms, white, "silver" and black. The mine is small. Due to the condition of the 4-wheel drive road to the mine, Stuart asks that anyone wanting to go to this mine meet him at his house by 8:30 am. If he thinks your vehicle can make it, you will be allowed to drive in, otherwise he will give you a ride in and back. The work day ends at 3:00 pm, but if you are on a really good pocket, he will allow you time to finish collecting it. However, the mine is closed by 5:00 pm, no matter what.

Crystal Hill, Montgomery County
Contact: Tim Hill
Address: Rt 1 Box 926, Glenwood, AR 71943
Phone: 870-356-4615

Website: www.eyesoftime.com/hillmine
Fees: $20 per adult, $10 per child (6-10)

All crystal they mine is sold wholesale, no retail shop. They will provide an experienced guide to show you where and how to dig crystal. Digging tools, buckets and gloves are available for sale or daily rental. You keep all the crystals you can dig. The owner says satisfaction guaranteed or your money back. Directions: Turn South off Hwy. 270 on Logan Gap Road. Drive approximately 2.6 miles and you will come to a cross road #177. Turn right, follow signs toward Collier Springs. Do not turn left toward Alamo. Proceed about one mile to concrete low-water bridge. Mine entrance is approximately 100 feet past the low-water bridge on the left. This is a primitive area with no facilities, services or telephones. Please stay away from the flagged area and machinery.

We were very pleased to meet with the owner and hear his philosophy of mining and tourism in the area. He told us that one year, after he was asked to be a judge at the annual Championship Quartz Dig in Mount Ida, he was disappointed with what the diggers found. Feeling that tourists

Those with walking disabilities will find the level ground at Fiddler's Ridge fairly easy to get around on.

were not getting a fair chance, he decided to open up a really good digging area, but at a higher price. The owner appears straightforward and no nonsense. He has business to do in the mine (he says 90% of digging quartz is moving dirt), but will have an employee take the time to show diggers where and how to work the mine. He asks what the digger is wanting, like singles, clusters, or jewelry points, and makes recommendations. Diggers can work on the dirt piles or on veins.

The short road to the mine was a tailpipe dragging road for our Honda, but we did make it. We did not spend the day digging, but saw evidence of the high quality material coming out of this mine. A great measure of respect for the owner is reflected in the number of times we heard the word 'integrity' from other miners and dealers who spoke of him.

Fiddler's Ridge, Montgomery County
Contact: Jim Fecho
Address: 3752 Hwy 270 E, Mount Ida, AR 71957 7 miles east of Mt. Ida
Phone: (870) 867-2127; e-mail fecho@ipa.net
Fee: $10 per adult. From the highway, the shop looks like it is down in a hole; you'll see the sign on the roof as you go by. At the shop, you get your permit and map to the mine, which is about 5 miles away. You get to dig through the tailings. Sandy ground is good for rainy days, and the flat areas are good for handicapped persons. I've collected here several times and always got what I thought was my fee-pay's worth. Some collectors report that they have had great success at this mine.

In the 1960's, this location was a mining claim of the Hot Springs Geology Club. There were one or two big piles of dirt to dig in, and some hand dug trenches then. The club dropped the lease, and when the Forest Service went to the contract system, this mine was parcel number one. There are no facilities at the mine.

Leatherhead, Montgomery County
Contact: Matt Price and David Huffman
Website: www.crystalmountaingems.com

The Leatherhead mine sold just as this book was going to press. The new owners are moving in from out of state and will be offering tourist digging. Contact them through their website.

This mine was first run by Tony Thacker, who is developing a new mine west of Mt. Ida. The new mine should be open early in the year 2000; you may make reservations with the Thackers at (870) 326-4871. Their retail shop is a small place at 3480 Hwy. 270 West, 12 miles west of Mt. Ida, and their the wholesale shop is five miles further west. You can tell the wholesale shop by the logo painted on the side of a tumbledown barn, a genuine Arkansas trademark. Tony is a bewhiskered fellow who took the time to talk to us while he worked on a batch of quartz at his processing yard.

Crystals from the Leatherhead mine are clear and lustrous.

We asked him about this curious name leatherhead, and he said it was a nickname bestowed on him by O.C. Carmack, a digger for Ocus Stanley in years past. Tony told us how he started out digging with a $9.95 grubbing hoe, and now he operates one of the biggest trackhoes in the area.

The Leatherhead mine yields mostly clear crystals, a few smokies, and phantoms. He pulled a small, carefully wrapped, beautiful phantom out of his pocket and showed us the phantom crystal pictured on page 98. We were impressed. He told us that he uses his big machinery to "face off" (or open up) a pocket of quartz, and those droppings go into the tailings pile, making it possible for diggers to find clusters as well as singles in the dirt piles. Our van full of cub scouts made it up the mine road deep in the mountains OK, and there is plenty of room even for big rigs at the top.

Telling us that a crystal doesn't have to be big to be pret-

A sample of crystals from the $1 table at Jim Coleman's shop. It is easy to get crystals this size and larger from Miller Mountain.

ty, Tony showed us the nice clusters pictured on page 72. He takes into account that the uniqueness, clarity, luster, and shine all influence the pricing of crystal. We were also amused to hear that a pocket of quartz could be sold anywhere from a case of beer to $10,000.

With new owners taking possession of this mine, we can't tell you what their policies of the site will be. Check with the new owners as to the status of collecting at this location.

Miller Mountain, Garland County
Operators: Bill and Lyla Norris; Owner: Jim Coleman
Address: PO Box 21, Jessieville, AR 71949
Phone: (501) 984-5752
Open every day; wholesale and retail.
Fee: $10 per person. Children under 10 years old are free when accompanied by a paying adult. Pay at the mine.
Restrooms, camping, showers, soft drinks & snacks

Diggers work piles of dirt that have been brought up from the mine pit of Miller Mountain in Jessieville.

available on site. Retail shop is Jim Coleman's on Hwy 7. Directions: Traveling north on Hwy 7 from Hot Springs, turn west on Hwy 298 by Castleberry's Store. Go 9.1 miles west on 298. Watch for a small sign that says Miller Mountain. Savall Circle is a dirt road to the left. Go down this road to the T junction and make a right turn at the sign. (This is one of those roads where you feel you're in the backwoods!) Take Bighole Road to the left and up the hill.

This mine is one of the major collecting sites in the Jessieville area and we like going here. You dig in piles of dirt brought up out of the open-pit mine. Large crystals are common. Singles and clusters can be found in the material brought daily from the mine to the collecting area. Their wholesale and retail shop may also have some unusual specimens. These may include attractive iridescent iron-oxide coated quartz and specimens of quartz with adularia or calcite in several forms or habits, and brookite. The calcite varies from distinct individual honey-colored crystals up to 2" in length, to crusts of small grey crystals and even cave-like formations.

Very rarely, needle-like aragonite has been found here. The brookite from here is different from that at Magnet Cove. Here it is long brown translucent blades with a peculiar metallic luster. It may be in the quartz or sticking through it. The old timers called these blades of brookite "steel splinters".

Robins Mining Company, Montgomery County
Contact: Mearl and Lorraine Robins
Address: PO Box 236 Mount Ida, AR 71957, Junction Hwy 270 & 27 South
Phone: (870) 867-2530
Website: www.robinsmining.com
Fee: $10 per adult. Open weekdays, Saturday by appointment. Shop open Monday through Friday 9am-5pm; wholesale and retail.

Collectors can expect to get crystals 3" to 4" at Robins, plus many smaller ones.

Another family-owned mine through the generations, Tim told us that their mine is a good place for beginning collectors to visit, because it is not as large or confusing as one of the big pits. He also said advanced collectors would be more satisfied digging at one of the other mines in the area. We must put in here that a quartz collector should certainly visit their shop in Mount Ida, as they have more nice quartz and fewer tourist doodads than any of the other shops in town. Their wholesale area is one of the most impressive. The fee pay mine is 4.5 miles from the shop, and you need to buy your permit first and get a map. Every digger gets to pick 5 pounds from their "crystal garden" outside (or $10 worth) before they go to the mine. It is easy to get garden and landscape rocks in the tailings. If the digger is skilled, he can get singles up to 4" and a few clusters. The owners say to bring your own tools, and a 6' prybar is recommended. They use heavy machinery to expose areas to work. The ground is not flat, but it is easy to get to. You can drive your

Clusters of Mount Ida Quartz have long, clear "candles" on the base. From the Robins Mining Company.

car to within 50 yards of the collecting area.

The wholesale area of their shop offers lots of singles and clusters in many price ranges. The very high quality ones are $100 per pound and they have tables of quartz from $5 per pound to $45 per pound. As is true with most places, they give discounts for volume buyers.

Ron Coleman Mining Company, Garland County
Contact: Lorene Teal
Address: PO Box 8219, Hot Springs, AR 71910 (Little Blakely Road, Jessieville)
Phone: (501) 984-5396; 1-800-291-4484
Website: www.omtp.com/colemans
Fee: $20 per adult, other prices for groups. Campground.

Perhaps the best known mine in the Ouachita Mountains, this open pit is the largest quartz mine and is one of the older active mines. From the shop, there is a great view of the mine and collecting area. At this location you don't get to dig on the veins, but when actively mining, material is brought out of the mine and dumped on a daily basis in the collecting area for people to pick through. They

Your digging hosts Lorene and Gene Teal at the Ron Coleman Mining Company. Dana stands by a museum piece. It's yours for the price of $85,000.

The Hot Springs Elderhostel classes collect crystal at the Ron Coleman Mine, Jessieville, Arkansas.

The open pit mine at Ron Coleman's. Truckloads of dirt are brought up from the bottom of the mine and dumped in a collecting area.

will provide you with a canvas tote bag and loan you a rebar tool for scratching around the ground. A nice plus is the crystal washing station, where you can hose the mud off your crystals.

Now in the third generation, the Coleman family operates over 100 acres of mining area, producing quartz for collectors and mining industrial-grade milky quartz as raw materials for man-made cultured crystals. Museum specimens of up to 6000 pounds have been recovered from here.

This mine has gone by many names and owners, including the Old Coleman Mine, Dierks #4, Blocker Lead #4, and Geomex.

The purity standard for quartz was first set from here by the National Bureau of Standards, and other mines in the Ouachita Mountains also meet the purity standard for industrial applications.

Sonny Stanley always finds the time to visit. Mike is holding some very large smokies Sonny had recently acquired from a mine in Paron.

Stanley Mine, Montgomery County
Contact: Sonny Stanley
Address: Pine Street off Hwy 270, PO Box 163, Mount Ida, AR 71957

Fisher Mountain is the site of the Stanley mine. Can you see the man?

We caught Sonny in his favorite chair under the oak tree. He says he has made many a friendship here.

Phone: (870) 867-3556 (day); (870) 867-3719 (night)
Website: www.mtidachamber.com/stanley
Fee: $4 per person.
Directions: To get to the shop, turn on South Pine Street in Mount Ida, across from the Family Dollar Store. Go up the black top road and take the left fork that says dead end. The sign reads Stanley Minerals and house.

The Stanley family operates the oldest continuously active mining claim/contract not only in the Mt. Ida region, but in the USA! Fisher Mountain mine is a long ridge with a steep well-maintained dirt road to the top. The mine is about 10 miles from the Stanley Crystal yard; you can buy your permits at either Stanley's or at Judy's Crystal's N Things at the turn off Hwy 270 at Logan Gap road (just west of the Mount Ida airport). Big rigs can make it to this site, but may require several back ups to get turned around at the end of the road at the summit. We do recommend Stanley mines if you have a lot of kids for a first-time collecting experience. It is quiet, open, and lots of loose single small points can be scratched out of the ground with a 3-tined garden tool. Bring your own food and water, there are no facilities at the collecting area.

Starfire Mines, Montgomery County
Contact: Charlie, Chris, or Shirley Burch
Address: 5403 Hwy 270 East, Mount Ida, AR 71957; at the Colonial Grocery & Motel
Phone: (870) 867-2431
Website: www.starfirecrystals.com

This mine was inactive during 1999, but the owners expect to start mining again in 2000. Shop is 25 miles west of Hot Springs, 11 miles east of Mount Ida. Wholesale and retail. Digging by appointment only, and you should call to confirm availability. The owners are very nice and will let you know the status of the collecting. You will find singles and clusters, and some tabular crystal. The crystals have a high luster, also fadens, phantoms, and burrs. The Starfire Mine is next to the County Line Mine in the Collier Creek area. There are no identifying signs, but the owners will give you a map and directions. There is a short rough road going to the mine, but our Honda made it okay, groaning up the hill. No facilities at the mine.

TNT, Montgomery County
(County Line Mine)
Contact: Darren Harris
Address: 2383 Hwy 270
Mount Ida, AR 71957
Phone: 870-867-5203
email: nika@alltel.net
Fee: $40 per adult

Crystal from the County Line mine.

You can dig right in the veins; plates and clusters are common. Specimens in the shop were very impressive. This mine, which is close to Fisher Mountain, has three different levels material is taken from. The owner says about thirty percent of production is double terminated. The shop sells mine run and wrapper baskets of quartz. Shop was formerly known as the Crystal Pyramid and changed owners in 1999. Wholesale and retail sales are made at the shop on Hwy 270.

Wegner Crystal Mines & Ranch, Montgomery County
Contact: Richard Wegner
Address: 82 Wegner Road, PO Box 205, Mt. Ida 71957
Phone: (501) 870-867-2309; 1-800-367-9888
Website: wegnercrystalmines.com
Fee: $20 Phantom Mine or $15 Crystal Forest Mine per person, minimum of 10 persons per group required. Children under 12 half price. Salted piles ($6) are easy for seniors and children. Camping, showers, and snacks are also available. Make reservations and check on holiday and seasonal schedule. Open 8-5; 7 days a week except holidays.

Located 6 miles south of Mount Ida off Hwy 27 on Owley Road. Blue signs show the way.

The Phantom mine produces blue phantoms and is their more popular and productive mine. The Crystal Forest mine has clear crystals. The owners will transport you to the mine and provide a guide. A large barn warehouses their large wholesale inventory and museum specimens.

Willis Mine, Saline County
Contact: Rickey Lakin
Address: 21250 Buffalo Road, Paron AR 72122
Phone: (501) 594-5228
Website: www.putpeel.net/peel/quartz.html
Fee: $5.00 for digging in the tailings, or $10.00 for digging on the veins. Children under 12 get in free with a paying adult. Call for information about group discounts. Hours are 8:30 am to 4:30 pm; earlier or later by appointment. Cars can drive to the mine site, which covers 10 acres.

This mine produced some of the largest and most attractive chlorite-phantom quartz in the Ouachita Mountain region. Parallel growths and tabular crystals also come from this site. This mine is noted for producing unusual shapes and sizes of crystals as well as chlorite included quartz. Great-grandfather J.T. Willis started digging quartz here in 1900 with his family. Some facilities at the mine.

Later chapters give general locations by counties, shown on the base map below.

If you look at topographic maps of Arkansas, you'll find at least seven other places called "Crystal Mountain" in Montgomery, Saline, Garland, Pulaski and Carroll counties. These aren't to be confused with the "Crystal Mountains" just south of Mount Ida.

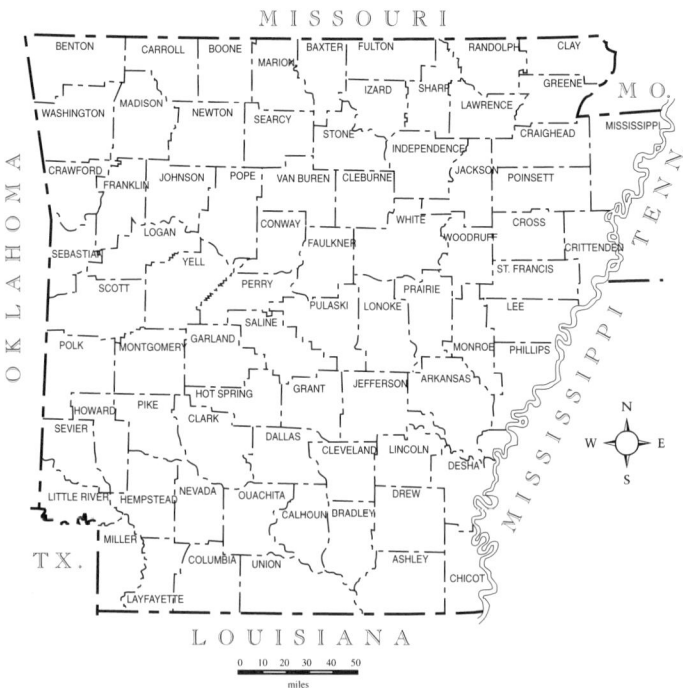

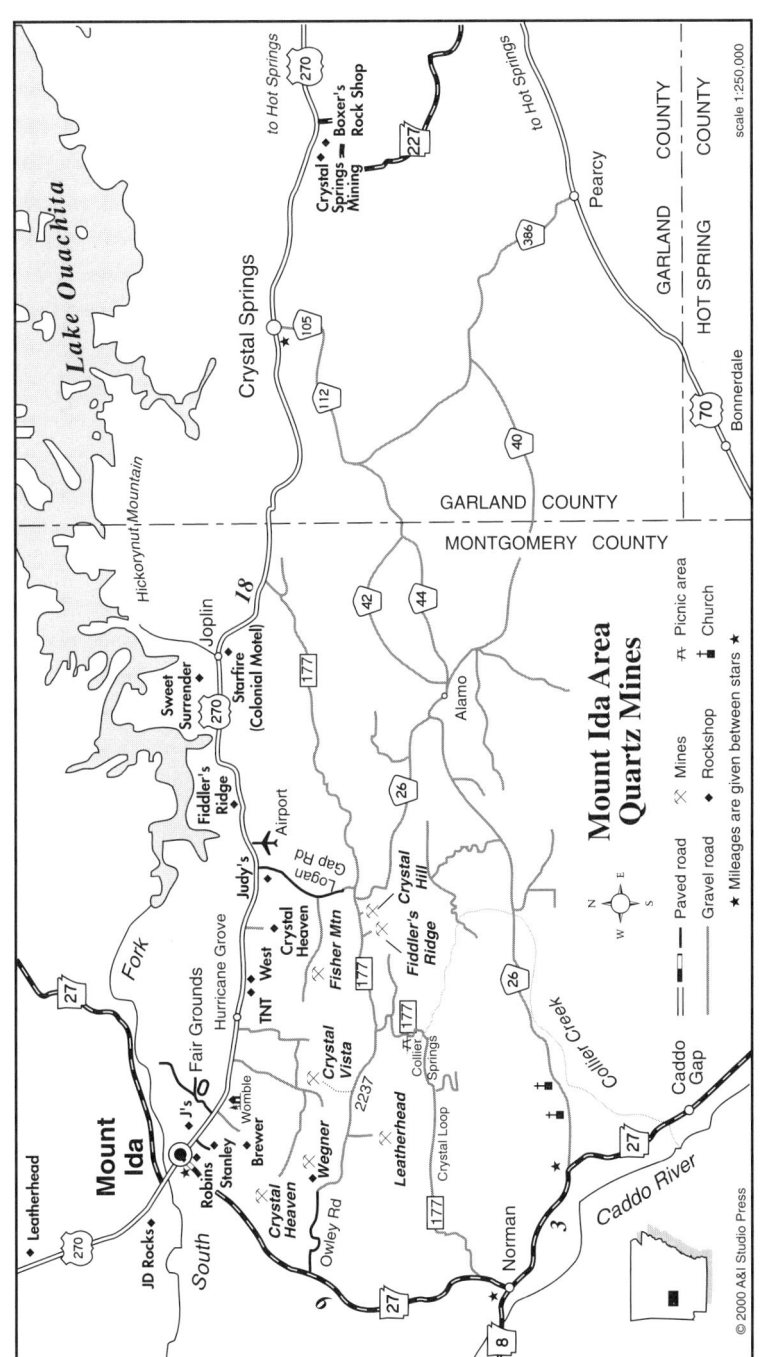

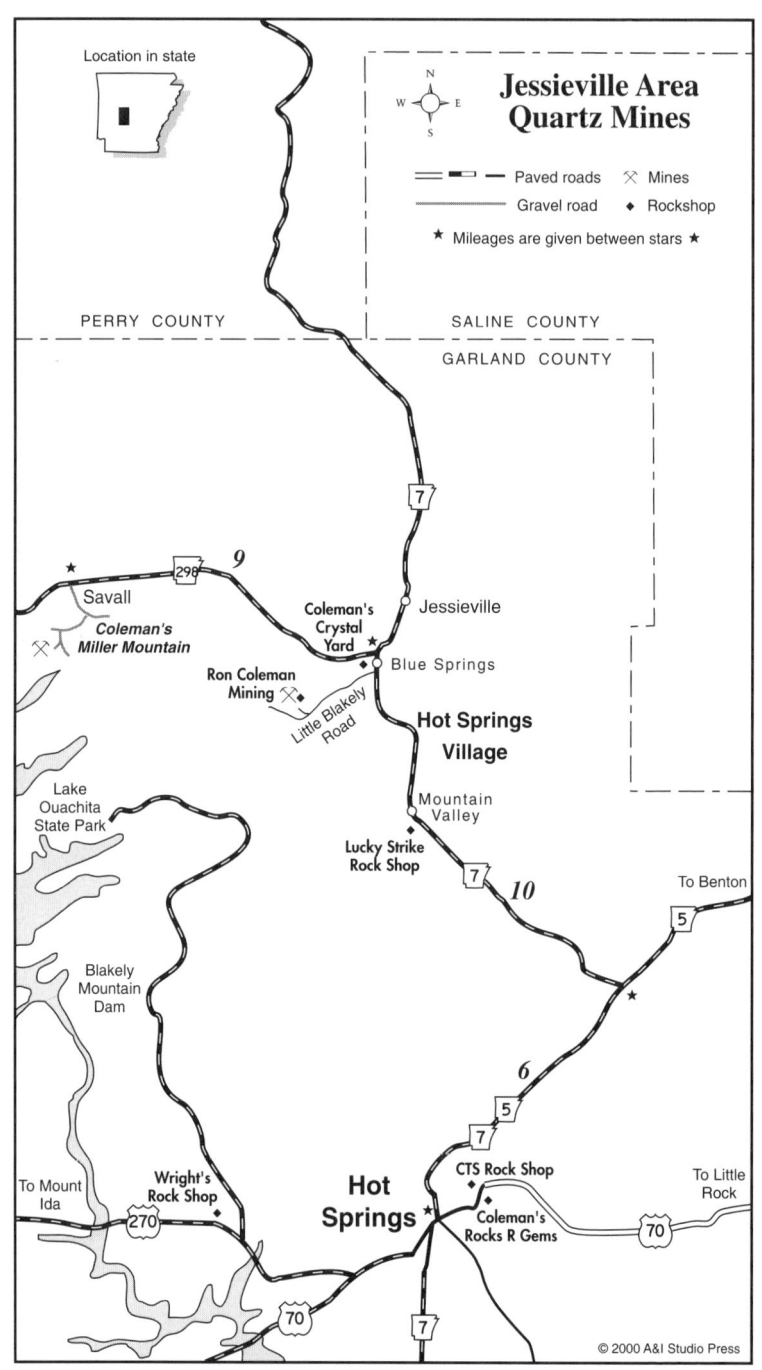

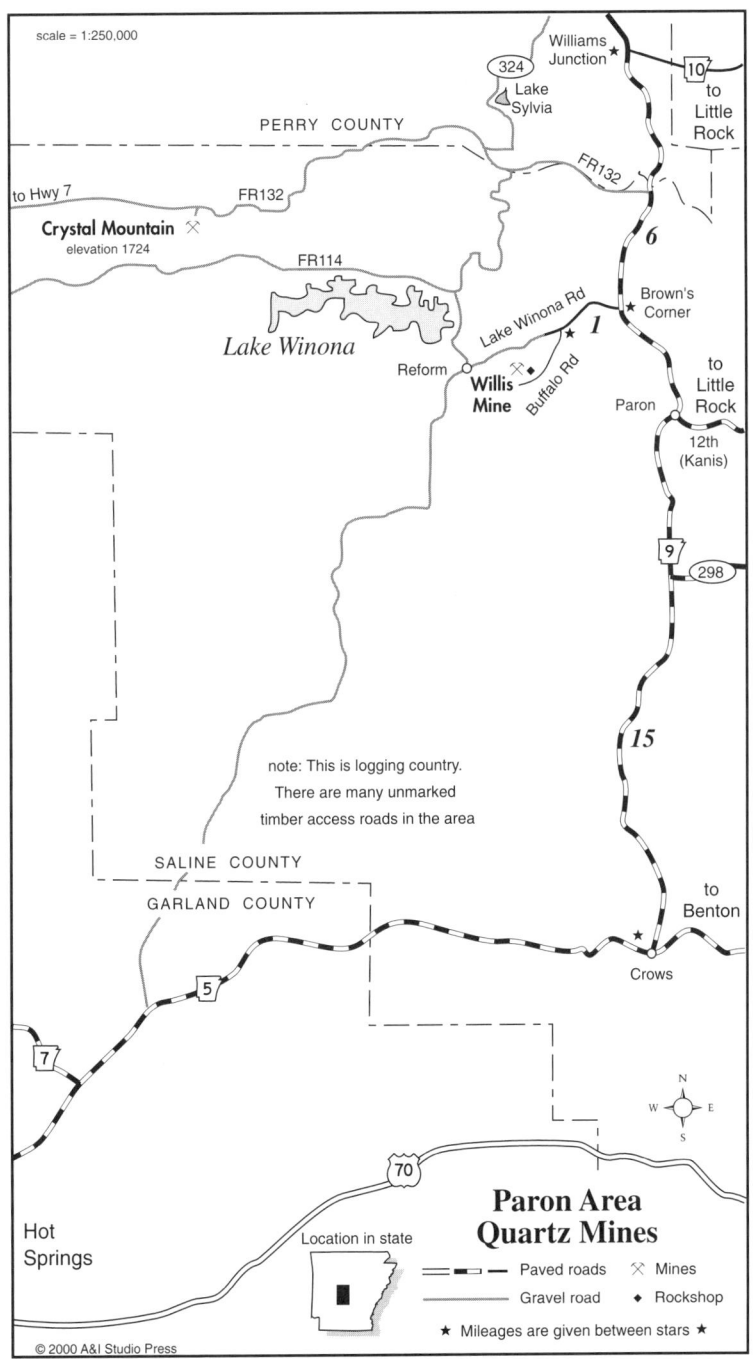

6. Kinds of Crystals

MANY VARIETIES of quartz can be collected in Arkansas. Some are prized for their beauty, while others are valued because of their oddity. Here are the kinds you will find when you dig, or look in shops and collections.

Known as either "points" or "singles", this type of crystal is the most common we find. The "classic" shape of a single crystal is also called "prismatic" quartz, recognizing

Single points and clusters from the Mount Ida area.

the straight sides and base. Long slender singles are called "candles" if they are six times longer than wide.

Clusters can be small with only two points, or large and massive with hundreds of points. Classic clear crystal is also known as rock crystal to differentiate it from other kinds of quartz, such as milky quartz, rose quartz or amethyst.

Bubble/fluid inclusion

Captured inside a crystal, water and air form a small cavity. These crystals are very interesting because when tilted, the bubble can be seen to move. Unfortunately, photographs can't show these features to do justice to them.

Cactus quartz

The growth in the long direction of individual crystals is so rapid that the end faces are not filled in on the crystal aggregates to form a "single" crystal. Tiny faces terminate each point, all in the same orientation. The effect is like a

Multiple terminated growth on a growth form called "cactus quartz" by local collectors, Big Cedar Bluff, Saline County.

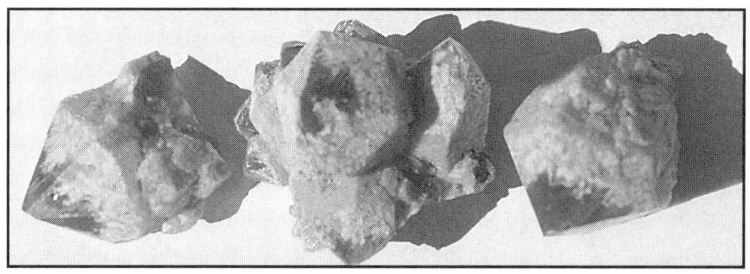

Coontail quartz from the Runyan quartz mine (now closed), Hot Spring County, AR.

cactus with stickers going out in all directions.

Coontail

This peculiar stubby grey and black smoky quartz came from north of Magnet Cove in Hot Spring county. When cut at a right angle to the point these crystals display alternating zones of milky and smoky quartz, like the rings on the tail of a raccoon. These crystals rarely get over 1 1/2 inches across and 2 inches long. Undamaged and well-cleaned specimens are hard to find anymore since the mine has been

Cubic quartz from Ron Coleman mine (left) and Willis mine (right).

closed and unworked for years.

Cubic quartz

Distorted double terminated growth forms are called "cubic" quartz by old time collectors. It has very small to non-existent prism faces. These crystals look like just the point parts of the crystal without the long straight sides.

Curved crystals

Growth of subparallel individual crystals may produce a curved crystal specimen. All the individual crystals in a specimen are straight, but the overall effect of offset growth is a curved appearance. However, sometimes the curving is due to multiple period of breakage and rehealing.

Double terminated quartz

These crystals have a point on each end. Some are short and stubby, making them look like the herkimer diamond

Curved crystal from Ron Coleman mine, Garland County.

A variety of double terminated crystals, clockwise from upper left: water bubble quartz - Glenwood, Montgomery Co.; smoky - Magnet Cove, Hot Spring Co.; black inclusion - Collier Creek mine, Montgomery Co.; two prismatic crystals - Jeffrey quarry, Pulaski Co.; large double with small attached double - Ron Coleman mine, Garland Co.; two herkimer diamonds, Herkimer Co., N.Y.; smoky water bubble crystal - north of Lake Greeson, Pike Co.; smoky from Smoky Crystal mine, Garland Co.

Drusy quartz coating sphalerite with smithsonite from Monte Cristo mine, Rush, Marion County.

crystals of New York, others are long and slender. Most of these crystals will have a milky zone in the crystal. This zone is where the crystal was broken from the vein wall and continued to grow, resulting in repair of the break and growth of a point on that end also.

Drusy quartz

Drusy quartz is present as coatings of uniformly small crystals lining vein walls or pockets. In the Ouachita Mountain region, drusy crystals may also encrust larger quartz crystals. In the case of quartz mineralization in silicified carbonate rocks, as in north Arkansas, drusy quartz may coat other minerals, like sphalerite or galena, or encrust pockets in the host rock.

Faden

Faden is a German word for "string" In this instance, slender strings of quartz grew across open fractures, then other quartz crystal grew in preferred orientation on the pre-existing strings, often as parallel groups. The key point of faden quartz is the milky line that crosses the crystal.

Crystal from Jeffrey quarry, Pulaski County, showing a faden line.

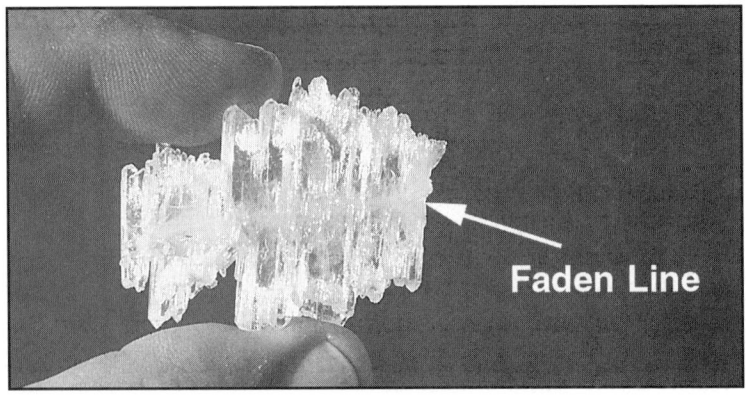

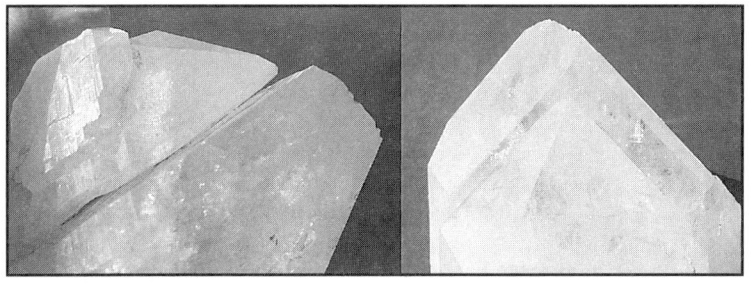
Faulted crystal, front and side view, from Ron Coleman mine.

Faulted and fractured crystals

Faulted crystals are a little different than fractured crystals. Many crystals get fractured and reheal so the fracturing does not reach the surface of the crystal. This is one way "rainbow" reflections are generated. Faulting of a crystal actually offsets the two or more pieces, therefore the two parts might still be joined to each other due to later growth, or might be recovered as two separate partial crystals, sometimes with regrowth on each broken portion.

Jeffrey solution quartz

The Jeffrey quarry, located in North Little Rock, AR, was the site of the discovery of solution quartz. At first, some collectors were certain the quartz was still growing in the original clay host, rectorite. But crystal growth at that time was impossible because the temperature was too cool and the original solution water had been flushed from the system.

This type of crystal got many collectors excited while this site was open. Burrs (crystals growing in all directions) formed in pockets of rectorite as many single crystals grew and interlocked together. Literally, millions of needle crystals were recovered from zones of gooey, vaseline-like rectorite-filled vugs in fractured quartz veins over the 15-year

Jeffrey quarry burrs and "haystacks", Pulaski County, Arkansas.

period that the site was accessible. Thousands of small clusters were available. The best quality clusters from this site were open, beautifully lustrous groups of transparent colorless crystals. Because slightly milky crystals were more typ-

Parallel growth of crystals, Jeffrey quarry.

ical, less common gemmy transparent burrs are most highly valued.

The site is closed to collecting, but specimens may be found for sale or trade.

Parallel growth

Parallel growth occurs when crystals which are similarly oriented crystallographically come in contact with each other. They may grow from a string, or faden, or they may grow from the matrix or vein wall. Often, parallel growth is seen where fluid movement appears to have been at a uniform rate and direction for a period of time. These conditions allow the crystals to have a preferred direction of growth and orientation in the vein.

Twins

Japanese twins have been scarce, but a notable number of Japanese twins have come from the Collier Creek mine and Fisher Mountain, both localities in Montgomery County, and even a few from the Old Coleman mine in Garland County.

Morphologically speaking, both Brazil twins and

Japanese twin crystals from Fisher Mountain mine, Montgomery County, AR.

A beautiful phantom from the Leatherhead mine near Mount Ida. This piece has a complete crystal point inside the larger point. From the collection of Tony Thacker.

dauphine twins from Arkansas as recognizable twins appear to be extremely rare. However, etchings of crystals from Arkansas indicates that they are very common. During World War II, most of the eye-clear quartz that was submitted for optical and electrical testing was twinned and failed to be usable. I think only about two percent was untwinned. The problem is that although it might look like a single untwinned crystal, it usually is not! Some specimens show a feature on the prism faces that look like zones or patches of slightly different reflection or luster. You can see the different lusters by turning the crystal slightly in good light and looking at the way light shines off the surface. These patches are bounded by curved to zigzaggy lines. This differing luster is an indication that the crystal is twinned.

Phantoms and inclusions

Phantoms are caused by a number of things that might happen while a crystal is growing. Any type of change, such as the chemistry of the water, growth interruption, or earth movement (structural adjustment) would have some effect on nearby quartz veins and the crystals forming in them. Sometimes just the type of host rock determines what type of inclusions may be present.

I'm using two terms, phantom and inclusion. A phantom

is a form of an older crystal seen inside the present crystal. It may be due to fine-grained mineral matter that was deposited on the earlier and smaller crystal, or a fine coating of bubbles as the pocket periodically dried out, or as ground up rock dust created by nearby faulting floated into the pocket and was deposited. An inclusion is simply any material captured and enclosed as the crystal grew.

In Arkansas quartz, inclusions can be many other minerals. Quartz has had a relatively lengthy time of growth, though episodic, when compared to other minerals. The following minerals and rock materials are often seen included in Arkansas quartz: adularia, chamosite (a variety of chlorite), cookeite, ankerite, calcite, pyrite-marcasite, quartz (as both sandstone grains and small crystals), and brookite. Other minerals have been noted as inclusions, but are somewhat rarer: cinnabar, stibnite, jamesonite, galena, rutile and clay minerals.

Shale inclusions (left), marcasite-pyrite inclusions (center) and the arrows point to cookeite inclusions in the crystal on the right.

Phantoms take on several forms. Perhaps the most attractive are those ghostly crystals containing essentially complete caps or terminations coated with some material to make the point display well. Often, this type consists of a fine coating of light-colored, almost transparent, mineral or tiny bubbles which formed on the point and were coated by the later deposition of clear quartz. Most often phantoms display only two or perhaps three of the prism or side faces of the crystal with a mineral or rock material coating them. Clouds of inclusions sometimes fill the early-formed crystal, which was then coated with colorless quartz. One mine in Saline County was named the White Cloud mine due to the white cloudy phantoms that were often recovered from it.

The materials most often composing the phantoms are chlorite, bubbles, and tiny shale particles. All of the so-called blue phantoms, black phantoms, "manganese" phantoms and "manganese" inclusions and "carbon" phantoms are the same stuff: finely divided particles of black shale, a relatively common host rock in the Ouachita Mountains. I have had many of these analyzed and the crystals are always aluminum-rich with no trace of manganese or carbon present, contrary to popular belief. Even some earlier geologists were mistaken on this problem by making guesses instead of having the chemistry run on the material.

The way phantoms grew is interesting, and the manner in which black or blue phantoms formed can be described. Most of these phantoms are of the type that have only a few body and point faces coated. Fine shale particles were carried into the forming quartz veins, suspended due to the flow of the silica-rich solutions. As this hot water moved through the pocket, the black shale particles settled out on the downstream side of the growing quartz crystals, due to the fluid eddies on the backside of the crystals. Like clay in a glass of water, any movement of the water kept the particles suspended. Since the water was continuously bathing the upstream faces of the crystal during its flow, that side was continually being swept clean of any particles. It is possible to use some phantom crystals to tell which direction the water was moving. When you look through a clear set of faces to see some phantoms developed on its backside, you are looking in the direction of the movement of the water as the crystal was growing.

I should also say something about chlorite. Little work has been done on included chlorite in Arkansas crystal, but wherever chlorite is present, the host rock is shale-rich. The iron, silica, and other elements necessary to form chlorite in the quartz veins are evidently derived from leaching of the nearby shaly units.

A note about dominant faces

Sometimes using a cluster of colorless crystals you can tell the direction of movement of water in the vein when they were growing. This is particularly true of flat plates or sections of vein wall when the crystals are all about equal size. A large majority of the crystals will have the set of dominant termination faces all oriented in the same direction. You can rotate the plate around in strong sunlight and all the faces will reflect light together. The dominant termination faces became dominant because they were on the leading edge (upstream side) of the water flow. So, when you can see the dominant faces, you are looking in the "downstream" direction of the water flow in the vein or pocket.

Smoky

The name smoky comes from the brown appearance of the crystal. It is caused by radiation of one form or another and aluminum substituting in the silica tetrahedral structure of the crystal. The stronger the radiation, the darker the

Top left a natural smoky from Smoky Crystal mine, top right a natural smoky from near Paron, and bottom is a natural smoky from Magnet Cove.

color becomes. Natural smoky crystals are a light to medium brown, and artificially irradiated quartz is dark brown to black. In Arkansas crystals, there are naturally occurring smoky crystals that have a light brown to grey color. Magnet Cove is notable for smoky quartz. Occasionally, natural smoky quartz in relatively large crystals are dug from isolated veins. These deposits rarely produce more than a few hundred pounds of smoky crystals.

Tabular

Tabular refers to the thin and platy shape of the crystal. Tabbies, as some collectors call them, are known from many deposits. Certain sites, principally in the Jackfork Sandstone, have a higher percentage of tabular crystal than

Tabular quartz from the Stand-on-your-head claim (left) and Jeffrey quarry (right).

other sites. Most of these deposits also have other minerals such as rectorite, cookeite, and even ankerite associated with them, showing that they are related to the Jeffrey quarry deposits. The Stand-on-your-head claim in Garland County, White Phantom mine and Big Cedar Bluff, both in Saline County, are locations that come to mind with a high percentage of tabulars relative to prismatic crystal. But other mines, including Starfire and Leatherhead in Montgomery County, also have considerable tabular crystal.

Quartz that has been altered by people

Several kinds of colored quartz are being sold in rock shops. One kind of quartz, which looks like the old-timey carnival glass, is called "aurora" quartz. The metallic rainbow colors are caused by a thin coating of titanium that is fused onto the crystal. Dealers will often send pieces of crystal that have a nice shape, but poor luster or clarity, to be aurora coated. "Aqua aura" is another trade name for quartz that has been treated with gold, yielding a transparent light blue color.

Colored quartz seems to go in fads, with certain colors or treatments being popular each year. "Polymer quartz" is a shop name for a colored plasticized coating in rainbow colors that appears to be airbrushed on the crystals.

Finally, some shops contain a large amount of irradiated quartz that looks like really nice smoky. This quartz has been exposed to radiation which turns it black.

When in doubt, ask if the specimen is natural or has been altered or enhanced.

And things that are not quartz

We are often asked, "What are the big, clear green and blue and colored minerals outside of all the rockshops?" It is slag glass, a waste-product of steel manufacturing, and somehow it seems to have found a home in Arkansas rock shops where it attracts tourists.

7. Other Quartz in Arkansas

THE BELT of rock quartz crystal across the core of the Ouachita Mountains is where the majority of prismatic colorless crystals formed, but there are quartz deposits of other types and varieties in the state that are also interesting to the collector.

Within the core area some interesting associated minerals are sometimes recovered along with quartz. We have already mentioned the brookite from Miller Mountain and the West mine. In 1999, at an undisclosed location north of Jessieville, rutilated quartz was discovered. Scattered golden hairs of rutile were captured by quartz crystals that formed between the wall rock and the main rock crystal veins. Most specimens require examination with a hand lens to see the rutile that is both included and on the surface of the quartz and in the surrounding iron oxide.

Amethyst, the purple to violet variety of quartz crystal, is present in veins and pockets at the Crater of Diamonds State Park. When deep plowed, sometimes pieces of these veins are brought to the surface, but good amethyst at this location is even more rare than the diamonds are. This amethyst has much the appearance of amethyst from the well known deposits of Brazil, even to having inclusions of the mineral goethite. Very minor veins of amethyst are present at other volcanic vents to the northeast of the Park. Amethyst-like drusy quartz was seen in narrow veins in the talc-soapstone pits in Saline County during active mining operations. Unfortunately, these mines are completely closed at present and are not accessible to collectors.

North Arkansas

Rock crystal is present in many of the abandoned lead-zinc mines in north Arkansas. The best known collecting location was the "cave" adjacent to the Monte Cristo mine near the ghost town of Rush in Marion County. A large opening, some 20 feet in diameter and 15 feet high was lined with thousands of short stubby clear-to-milky quartz crystals. They were difficult to collect because they were firmly attached to a silicified dolostone rock. Crystals sometimes enclosed chalcopyrite crystals and some of the quartz was coated with later formed calcite and/or aragonite. This site is now inside the boundaries of the Buffalo National River and is off limits to collecting, having been fenced and gated a few years ago. However, many other old mines exist in North Arkansas, most on private property. These would be sites to consider exploring for quartz and other minerals.

While on the subject of north Arkansas, a very few, but interesting, double terminated quartz crystals have been recovered from limestone-dolostone quarries near Batesville, Independence Co. The quartz crystals were perched on crusts of saddle-shaped tan dolomite. These water-clear crystals look like herkimer diamonds.

Skeletal and water bubble quartz

In the Ouachita Mountains next to the core area, skeletal quartz crystal may be found. These crystals go by many different names: skeletal, mud-inclusion, water-bubble, and negative crystal. The host rocks are younger than those of the core region, being the Stanley Shale (Mississippian) and the Atoka Formation (Pennsylvanian) as opposed to the Blakely or Crystal Mountain Sandstones (Ordovician). Most skeletal quartz has a smoky tint, even a wispy smoke pattern, to it. Due to its skeletal nature, this crystal often contains trapped water, visible especially through the terminal faces. They are rare finds and treasures to collect! The fluid-filled cavities sometimes contain bubbles you can see move when you turn the crystal, and, rarely, even particles of black shale or tiny crystals of sulfide minerals, that move around when the specimen is rotated. Often, secondary clays have penetrated pinholes located on the surface of the crystal faces and filled the cavities.

The pockets that these crystals formed in were originally calcite veins that weathered away leaving the crystals to be found loose in the soil or in clay. Double-terminated crystals from such pockets look like herkimer diamonds in form. I have seen this type of quartz at only one site in normal looking quartz veins. Sites from which skeletal crystal are recovered are scattered throughout the southern Ouachita Mountains, but notable localities include one location north of Glenwood near AR Hwy. 8, one north of Story on AR Hwy. 27 just across the Yell County-Montgomery County line, and one site near Russellville in the Atoka Formation that yielded large pieces and crystals. Two sites I visited that had skeletal crystal in place in the veins are now inaccessible. The Pigeon Roost barite mine in Montgomery County had veins of barytocalcite crosscutting shales of the Stanley Formation. In some pockets of crystallized barytocalcite were small smoky quartz crystals of this type. The other site was at the old Bird and Sons roofing granule pit near Caddo Gap in Montgomery County.

Vertical calcite veins that cut the Stanley Formation were exposed in the walls of this pit. A few pockets were exposed that contained smoky negative crystal quartz perched on the calcite. Unfortunately this pit is now filled with water. There are also isolated pockets of smoky skeletal crystal sometimes found in road ditches along the north side of Lake Greeson in Pike County.

Antimony and mercury districts

There are two old mining districts in the southern Ouachita Mountains, both in the Stanley Shale, that contain some interesting and collectable quartz. The oldest known of the two is the antimony district in northern Sevier County, near the community of Gillham. In this region, many old mines and prospects were opened during the search for antimony, a metal used as a lead hardener and flame retardant. These deposits are quartz-hosted, the veins pinching and swelling. Comb-like quartz is found on mine dumps and tailings piles. The most interesting from this area are those quartz crystals that contain inclusions of stibnite as irregular metallic blebs. Often the specimens are coated with powdery yellow antimony oxides, like stibiconite, so that the inclusions are not readily noticeable. You have to remove the secondary oxides to see them. Several mine dumps in the area have yielded good specimens. One site, the May mine on the eastern end of the district, yielded some jamesonite inclusions as steel-gray needles in small colorless double-terminated quartz crystals, and also a few ankerite and siderite inclusions.

The mercury district extends from northern Howard County eastward across Pike County to near Arkadelphia in Clark County, but the bulk of the old mines and prospects are near Lake Greeson in northern Pike County. The Parnell Hill mine, now flooded by the lake, yielded a large number of small double-terminated quartz crystals with wire-like inclusions of stibnite. Small, bright-red cinnabar crystals are perched on the stibnite in some of these crystals.

Although one needs a hand lens to examine these inclusions, they are spectacular to see! Many of these crystal specimens were donated by the mine operator's family to the Arkansas State Mineral Collection, held by the Arkansas Geological Commission. I have rarely seen examples of these types of inclusion crystals for sale or in private collections.

Overcoat quartz

Milky quartz is widely distributed in the Ouachita Mountain region with veins known in essentially every sedimentary rock formation. Rarely are attractive specimens of milk-white crystals present, but some do occur.

Iron oxide coating on overcoat quartz from Collier Creek mine. Often sold as "manganese" coated quartz, but not correctly so. Manganese minerals associated with quartz are always coal black and have no iridescence. You can tell them apart because manganese oxide will streak test black; iron oxide streaks red or brown.

Years ago, Dr. Hugh D. Miser—a native Arkansas geologist who worked for the USGS—had shown a location in Polk County to a co-worker of mine, Ben Clardy. In the mid-1970's I was working with Ben on a project concerning antimony-bearing veins in Sevier County. One day Ben suggested we drive up Highway 71 and see if we could still find this place Dr. Miser had shown him. After spending time looking over the road cuts, we found the 18-inch wide vein. Using a pry bay, we opened a pocket and collected several iron-stained clusters. When I brought them home, I cleaned them to remove the heavy iron-oxide coatings and had quite a surprise! A creamy white layer of milky quartz coated colorless rock crystal, shown in the photo on the next page. Having never seen anything like these crystals from Arkansas, I termed them "overcoat quartz," indicating that the milky quartz was coating earlier quartz crystals. Later in the year my friend Art Smith told me those crystals just *had* to have come from Colorado. He was incredulous that I had collected them myself! However, he now has this site listed in his book *Collecting Arkansas Minerals - A Reference and Guide* (1996). Art calls them "snow crystals", but I prefer "overcoat quartz" because it describes how the crystals appear and formed.

In the early 1990's I was visiting West's crystal shop in Mount Ida and came across a table of quartz that looked identical to that which I had collected at Wickes. When I asked about the location, I was told that they had recently mined it from a site at Don Burrow's Collier Creek mine. I bought several hand specimens, and after cleaning them, they had a similar appearing coating of milky quartz over colorless crystal. So there are at least two sites in Arkansas that have yielded attractive specimens of white milky quartz crystals that are significantly different than the typical "run-of-the-mill" milky quartz crystals seen in so many of the state's quartz mines.

Overcoat quartz, or milky quartz coating clear quartz, makes this piece a kind of reverse phantom. From south of Wickes, Polk County.

In the previous chapter we listed the quartz mines that are open to public digging. By no means are those the only mines in the Crystal Mountains or the Ouachitas. Many private mining operations work numerous other sites and sell quartz to rock shops locally, around the nation, and to overseas buyers. Occasionally the miners will come across some unusual quartz or associated minerals, and these can be found in displays or for sale.

Quartz is the second most
common mineral. What's the first?
Read on!

8. Mineralogy of Quartz

Quartz has a unique structure and so do other minerals. That, in combination with the chemistry, is what makes a mineral a mineral. Here we'll talk about the properties of crystalline quartz, and finish up with a list of what we know about quartz vein formation in the Ouachita Mountains of Arkansas.

Let's begin with some basics: To meet the definition of a mineral, quartz must be composed of an orderly arrangement of certain elements so we may describe its internal structure and present its chemistry by a formula: SiO_2, "silicon dioxide" or silica.

Feldspar is the most common mineral in the earth's crust, but any mineralogist would agree with me when I say that quartz is the most diverse species in terms of varieties, shapes and forms for a single mineral. The feldspars or the pyroxenes and amphiboles include a whole host of minerals with similar structural characteristics, but variable chemistry. Quartz certainly has the most COLLECTABLE varieties of any single species. This includes the crystalline varieties of rock crystal, amethyst, smoky and citrine, and the cryptocrystalline varieties of chalcedony, agate, chert, flint, chrysoprase and jasper.

Here is a hand-sized piece of metallic silicon. This metal combined with oxygen makes quartz crystals: silicon dioxide. Silicon doesn't occur naturally like this piece grown in the laboratory, but is a combined component in a large part of the earth's crust. Isn't chemistry amazing?

Silica

We know that quartz is the low-temperature stable form of silicon dioxide. Several other forms of silica exist at higher temperatures and pressures. Quartz forms over a temperature range, the upper limit of which is 867°C at one atmosphere of pressure. If you think of inside the earth as a giant pressure cooker, different things happen at the higher temperatures and pressures than we see on the surface around us.

To compare the formation of quartz with something you can see here on the surface, the temperatures of an industrial blast furnace where iron is processed is about 400°F (200°C) at the top of the furnace, and near the bottom it is about 3,000°F (1,650°C) or higher. So for making crystals,

The kitchen chemistry recipe for silica syrup and crystals:

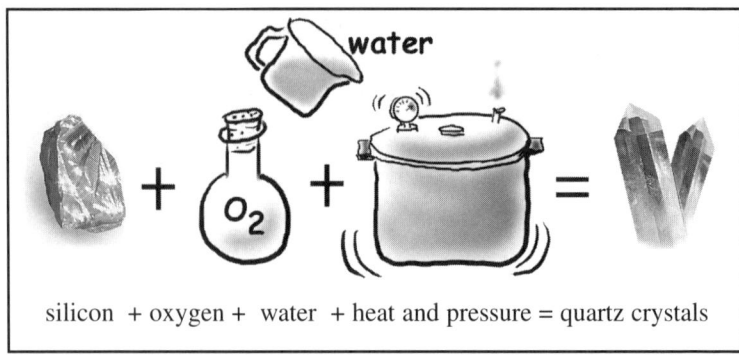

silicon + oxygen + water + heat and pressure = quartz crystals

we see the conditions much hotter than a pizza oven (450°F), but less than a steel mill.

Quartz crystals won't melt unless you chunk them into a foundry, but they can fracture due to a few degrees of rapid temperature change (see Chapter 4).

Alpha and beta quartz

Mineralogists get into details and describe quartz crystal as alpha-quartz (low quartz) or beta-quartz (high quartz). Alpha-quartz forms at temperatures lower than 573°C at one atmosphere pressure, where beta-quartz forms at the temperatures above 573°C and lower than 870°C at the same pressure. If the pressure increases, so may the temperature of formation of both alpha- and beta-quartz. For example, at about 2 miles deep in the earth, alpha-quartz may form at as high as around 600°C and beta-quartz at over 1000°C. These conditions may exist in our present

Faces on the point of a crystal will generally have differing sizes, making two sets of three faces. One is usually dominant. Commonly, crystals have six edges on the dominant termination face, but can be found with three to eight edges.

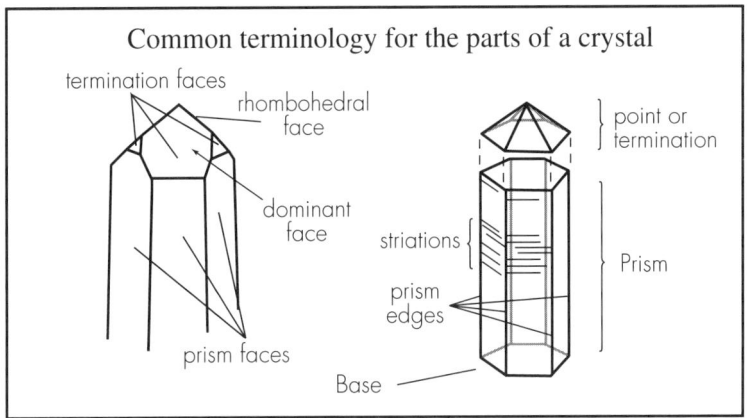

world today at the margins of the continental plates in subduction zones or at a depth of about 2 miles below where you happen be reading this book.

Beta-quartz is relatively uncommon, most occurrences being confined to rhyolite lava flows where the mineral "froze" in the rapidly cooled rock. Beta-quartz from rhyolitic lava appears as small equidimensional crystals floating in the fine-grained matrix. The beta- to alpha-quartz transformation occurs instantaneously at 573°C and requires little energy to "kink" the hexagonal lattice to a trigonal configuration. The external beta form is retained, but with an alpha internal structure.

All the quartz from Arkansas is alpha-quartz—so we'll simply call it quartz. Studies on Arkansas crystal indicate a range of temperatures of formation, from as low as around 200°C to about 265°C. This is well above the boiling point of water at atmospheric pressure, so there existed confining pressure due to depth. Perhaps as much as 2 miles of sediment and rock overlaid the principal quartz-bearing formations when the veins were forming. Today, these formations have been exposed by over 200 million years of erosion.

Physical properties of quartz

Textbooks describe quartz as a mineral with a nonmetallic luster, variable color, hardness of 7, white streak, specific gravity of 2.65 and belonging to the hexagonal crystal system and trigonal subsystem. Quartz also has several unique physical properties:

Cleavage Although quartz has the most cleavage directions of any mineral (7), it is listed in many mineralogy texts as having none! The reason is that quartz cleavage is rarely seen in nature. The cleavages that are known have all been discovered in the laboratory by either electrical or thermal shock. Since cleavage is not easily induced in quartz by

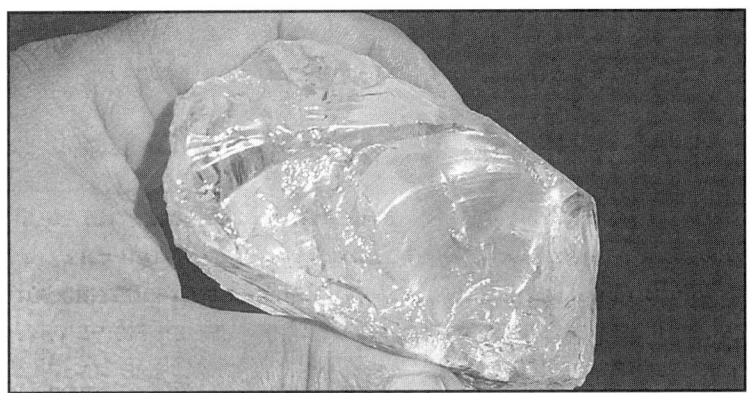

The conchoidal fracture of quartz is easily recognized.

impact, we may say the mineral typically has no cleavage.

Fracture Fracture is simply the manner in which a mineral breaks when cleavage is not well developed. Quartz has a fracture called conchoidal, meaning shell-like. The mineral fractures equally well in any direction. If you look at the broken edge of a piece of glass, you will see conchoidal fracture. This property was recognized by early man as a very useful one. If you are a flint-knapper, with practice you can learn how to control conchoidal fracturing. (We have an interesting modern arrowhead chipped from a broken Pepsi bottle!) Once prehistoric man mastered the technique of making projectile points using chipping he gained a degree of independence. He could simply carry some basic materials with him, and as he needed them, he could stop and make more tools. However, flint and chert, both microcrystalline varieties of quartz, are more readily available and easier to chip than rock crystal. Early oriental artisans carefully worked crystals into spheres by fracturing large blocks of rock crystal to attain a roughly rounded shape, then grinding them in a trough with water and sand.

Hardness Due to its internal structure, quartz has equal hardness in all directions. At 7, it also is the hardest of any of the *common* minerals on the Mohs hardness scale. Since it has no easily induced cleavage and is equally hard in all directions, the mineral is not worn away very rapidly during transport. This hardness explains why it is the most common broken-down (detrital) mineral in sediments, and why the beaches are sandy! Remember Mohs scale of hardness? Here's the jingle: The Girl Could Flirt And Flirt Quickly Though Connie Didn't (Talc, Gypsum, Calcite, Fluorite, Apatite, Feldspar, Quartz, Topaz, Corundum, Diamond).

Doesn't dissolve easily Quartz won't dissolve in most fluids. Note that I said in most fluids, like normal ground water. However, in carbonate-rich water and in very salty water with a lot of chlorine and sodium, quartz is somewhat soluble, especially if the water is heated. Quartz from the Ouachita Mountains formed from hot water, expelled from some depth during and shortly after the mountain building processes were active.

Cool to touch Quartz is a poor conductor of heat. Ancient peoples were well aware of this property. Objects and spheres carved from quartz always feel cool when touched or held, even in the heat of the day.

Piezoelectric property Several minerals, including tourmaline and sphalerite, exhibit this effect. When you alternately apply and release pressure on a quartz crystal, during the pressure changes a small amount of electricity is made. So by applying cyclic pressure, a current may be generated. Vice versa, when electricity is applied to a crystal, the internal structure vibrates. This principle is involved in the manufacture of highly accurate quartz watches and quartz tuners on stereo systems.

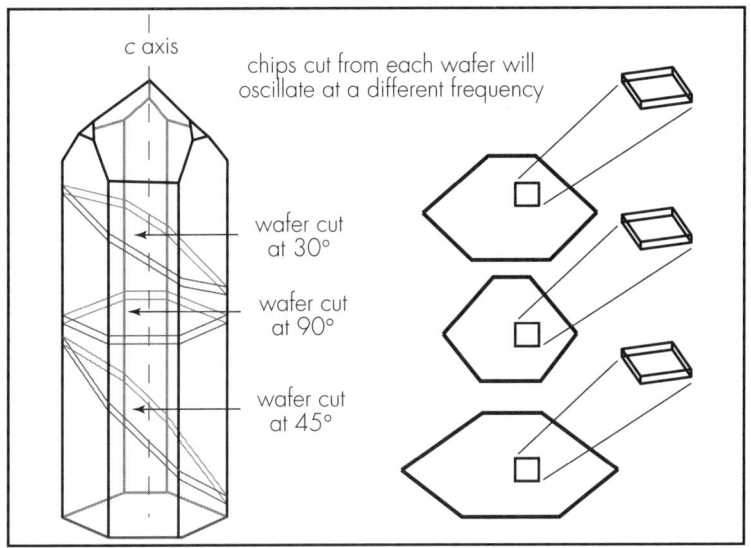

During World War II, very pure, untwinned pieces of quartz were in high demand for radio oscillators. The term crystal in CB radios was first used in the electronics industry for quartz crystal wafers, although now substitutes have replaced quartz. By cutting the wafer at a certain angle to its c crystallographic axis, we can control the frequency of the vibration. The original crystals in CB radios were cut from wafers of quartz, each having a specific frequency they vibrated at. This angle was the key to different channels on the radio. Early CB radios had eight wafers for eight channels.

Triboluminescence Luminescence is the emission of light without heat—fireflies are an example of luminescence. Triboluminescence is light that is produced by pressure, friction, or mechanical shock. It may be readily demonstrated with two hand-sized milky quartz crystals in a darkened room. Simply take the long edge of one crystal and rub it back and forth on the side of the other crystal. You may also rub two surfaces together, but you get more light using a prism edge against a prism face. This eerie light makes a good demonstration in a darkened classroom!

There have been numerous instances of lights reported during earthquakes, apparently coming from the ground. These lights are a special type of luminescence termed *seismoluminescence*. It is due to the release of light by countless quartz sand grains rubbing against each other during an earthquake. Seismoluminescence is mechanically generated during the movement of L (surface) waves.

Cathodeluminescence Cathodeluminescence is a distinctive visible color that is emitted by bombarding a small piece of quartz with cathode rays. This must be done in a vacuum to best see the visible color. Trace elements influence the cathodeluminescent color of the mineral.

Star of the c axis Asterism is not present in all quartz specimens. To see this property exhibited the specimen is best cut into a sphere or at least a high domed cabochon. Alpha-quartz that forms at higher temperatures may have other

Resembling a star sapphire, the six-rayed asterism of quartz is seen with light shining through its c axis.

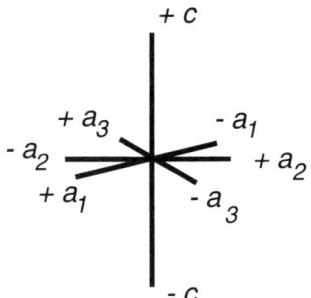

In this diagram of crystallographic axes, the c axis is the vertical one. The star in the photo is produced by shining a bright light from the bottom of the c axis, and the asterism is seen on the top.

chemical compounds that are "dissolved" in the structure. As the mineral cools, the dissolved material exsolves (unmixes) out of the quartz structure into discrete mineral particles. In the case of asteriated quartz, the dissolved material is thought to be very small amounts of titanium dioxide, TiO_2. When it exsolves, it becomes oriented along the three principal *a* crystallographic directions. These lie in a plane at a right angle to the *c* axis, and each of the 3 *a* axes are at 120 degrees to each other. When light shines on a sphere or is reflected back through a sphere of quartz that exhibits asterism, a sharp 6-rayed star appears when the sphere is properly oriented. You will need to rotate the sphere around until you see the star, then you are viewing down the *c* axis. Asterism is present in many minerals, particularly gemstones of the hexagonal system, like ruby or sapphire.

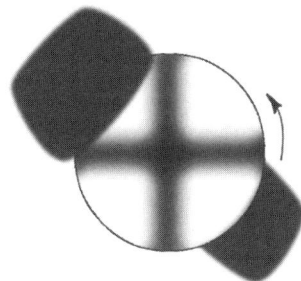

Optic axis Quartz is optically uniaxial, so if you place a quartz sphere between two crossed polarizer plates and rotate the sphere you may be able to find a uniaxial optic axis figure. It

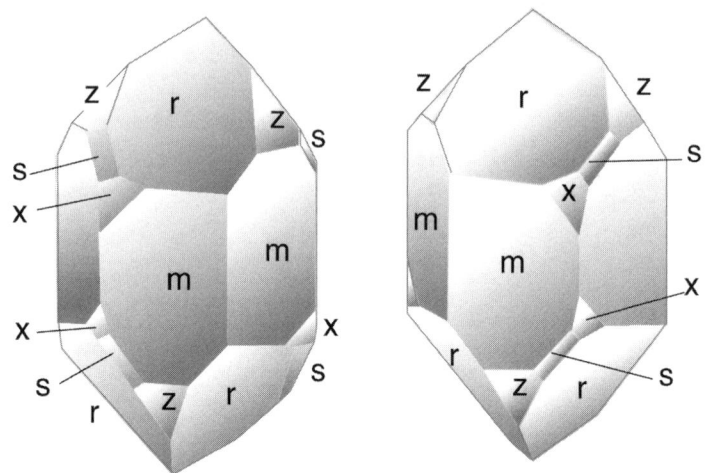

Left-hand and right-hand quartz, showing the letter symbols of different faces.

looks like a large fuzzy cross, and due to the thickness of the sphere would have multiple color bands coming out from the center. Two lenses from an old pair of polarized sunglasses can be used to check this property. Rotate one of the two lenses until the field goes black or dark when you look through both lens together, then insert the sphere between them.

Crystal structure The basic building blocks of a quartz crystal are silica tetrahedra. In quartz, these tetrahedra are linked corner to corner to build up the crystal. During this linking, or bonding, the overall structure may twist to the left or right as we view the crystal vertically along the *c* axis. Because the crystal's structure twists, we call this property enantiomorphism, a fancy term for right- or left-handedness.

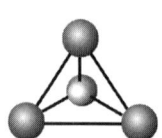

Tetrahedron shape

The term simply means that their respective structures are mirror images of each other. With close examination of a quartz crystal and a knowledge of what growth faces are present, one may determine if the crystal is left- or right-handed.

A silica tetrahedra consists of a single silicon atom linked to 4 equally spaced oxygen atoms, a kind of molecular Tinker Toy. The tetrahedra are linked together in a ring-like manner in layers. The tetrahedra alternate in the structure—one with the point up, the next with the point down. These linked rings spiral around the c crystallographic axis in either a clockwise or a counter clockwise manner. This structure was discovered by 19th century investigators observing the rotary power of various crystalline materials on light, long before the advent of X-ray diffraction analysis. The Curies were two early investigators in this field.

Specimen grade quartz

Collectors gather this material, which is also known by the names of water-clear, eye-clear, radio-grade or oscillator-grade quartz, and rock crystal. The term "rock crystal" is used for the colorless, clear variety of quartz that has made Arkansas famous among mineral collectors. The early mineralogist Dana used this term years ago and I like it to distinguish the colorless crystalline quartz from other types.

Industrial grade quartz

Quartz used in the electronics, computer and other high tech industries must be of a very high chemical purity, generally less than 20 ppm aluminum, 3 ppm iron, 10 ppm sodium (G. Coleman, 1987). While either the massive or the crystal form of quartz may meet the high purity standards, it is typically only the massive, milky, translucent form that is actually mined for industrial purposes. Often referred to as "bull quartz", industrial grade quartz is crushed and processed to produce various products including man-made synthetic or "cultured" quartz crystals. Other products such as computer chip wafers are cut or otherwise processed from the commercially grown crystals.

Now we can deduce...
Knowing about the physical properties of quartz can tell us something about the mineral's formation in veins. Having seen many veins in the field, we can use the physical properties and the field evidence to make the following statements about the conditions that existed at the time of quartz growth in the Ouachita Mountains:

- Growth took place at some significant depth (1 - 2 miles).
- The quartz grew from hot water solutions (over 200 degrees C).
- The water was rich in dissolved silica and was salty.
- During crystal growth, earth movement and vein adjustment were both active.
- Sandstone beds were favored for the formation of a higher percentage of clear quartz crystal because they were more fractured and provided better nucleation sites for quartz to begin growth.
- There were several periods of crystal growth in the veins over time.
- Temperature generally decreased during the period of crystal growth.
- Quartz veins may be either simple or complex in form, depending on the local geologic history.
- Quartz veins are more numerous in the tightly folded portions of the sedimentary beds than other areas.
- Veins containing rock crystal may extend for significant depth if a favorable host rock is present.

Selected references:

Anthony, J. A., Bideaux, R. A., Bladh, K. W., and Nichols, M. C., 1995, Handbook of Mineralogy, Vol. II - Silica, Silicates (part 2): Mineral Data Publishing, Tucson, AZ.

Dake, H. C., Fleener, F. L., and Wilson, B. H., 1938, Quartz family minerals, a handbook for the mineral collector: McGraw-Hill Book Company, Inc., New York.

Engels, A. E. J., 1952, Quartz crystal deposits of western Arkansas *in* Contributions to Economic Geology, 1951, U. S. Geological Survey Bulletin 973-E, Washington.

Howard, J. M. and Stone, C. G., 1988, Quartz Crystal Deposits of Western Arkansas *in* Proceedings of the 22nd Forum on the Geology of Industrial Minerals, G. W. Colton, ed., Arkansas Geological Commission Miscellaneous Publication 21.

Miser, H. D. and Milton, C., 1964, Quartz, rectorite, and cookeite from the Jeffrey quarry, near North Little Rock, Pulaski County, Arkansas: Arkansas Geological Commission Bulletin 21.

Newsom, Gene, 1978, The Jeffrey quarry: Mineralogical Record, vol. 9, no. 2, p. 75-79.

Ober, J. A., 1994, Electronic and Optical Materials *in* Industrial Rocks and Minerals, 6th edition, D. D. Carr, Sr. Ed., SMME, Braun-Brumfield, Inc., Ann Arbor, MI.

Palache, C., Berman, H., and Frondel, C., 1962, Dana's System of Mineralogy, Vol. III - Silica Minerals: John Wiley and Sons, University of Chicago.

Sosman, R. B., 1927, The properties of silica, an introduction to the properties of substances in the solid non-conducting state: American Chemical Society Monograph Series Number 37, The Chemical Catalog Company, Inc., J. J. Little and Ives Co., New York.

9. Frequently Asked Questions

SEVERAL QUESTIONS come to us fairly often. Beginning with the most frequently asked question, we'll try to answer the common concerns of diggers.

Where is the best place to go dig?
We can't completely answer this question... It depends on what you are wanting to do. The best crystals are at the fee pay mines. If you have kids who would be happy to find just a few small crystals, the Forest Service land is fine, or Stanley's would be good. If you need a mine with all conveniences close by, Colemans. For flat ground to walk on, Colemans or Fiddler's Ridge. All the fee-pay mines have good crystal to find. There is a fairly good correlation with the price of admission to the quality of crystal you can expect. The old "you get what you pay for" cliche applies here, but still it depends on the kind of production they are getting at the mine the day you visit. Some days are terrific, other days are spent moving dirt looking for the next big find. That's why we've presented information on all the mines to help you make your choice.

When is the best time to go crystal digging?

The joke about Arkansas weather is that if you wait 15 minutes, it will change. October is our favorite season, the weather is usually cool with fewer wet days than the springtime. The tourist season starts in March, around spring break, and goes pretty strong until school starts. There is a lull after Labor Day, making the end of September through Thanksgiving a good time to visit.

Each of the four distinct seasons has its merits. Spring is a lovely time to come here, the dogwood trees bloom at Easter and are a memorable site out in the woods. Daytime can be nice for working on your suntan, but there is also a chance of taking shelter from a storm during March, April and May. Arkansas has an average of 20 tornadoes a year, most of them in April. Be sure and keep up with the forecasts.

Most of the tourist activities are on a well-beaten path, but if you venture off on your own, as the foliage thickens and the weather warms, come prepared for sticker vines, rip briars, poison ivy, insects, snakes, etc. Early spring can provide some great collecting, especially before the leaves come back on the trees. Its cool enough for heavy digging and no snakes or bugs are active.

Summer gets so hot that I once saw a digger bar jump up and hop into the shade all by itself! Seriously, if you're digging on an outcrop, you'll need to make provisions for shade and rest. Pack your cooler with plenty of water and sports-type drinks; heat-related distress can happen quickly to a digger in the summer heat. Summer thunderstorms in the afternoon are common, adding to the normally high humidity.

Fall is probably the best time to come to Arkansas. The weather is good and storms are not common. The colors of the trees, especially in the northwest part of the state, are beautiful, with Eureka Springs being one of the most popular destinations. If you are going to that area, make your reservations early. The leaves come off the trees starting in middle to late October, making it easier to to see where you are going in the woods, but many hunting seasons—with guns, not rock hammers—also coincide with fall. Geologists like to do their field work here in the late fall and early to middle winter because leaves are off the trees (visibility is better) and the snakes, chiggers, mosquitoes, and other crawley critters are snuggled up for a while.

In early winter you may be rewarded with some unseasonably warm weather that would be great for rockhounding. January and February are the coldest months. Snow is rare and we are largely unprepared for it when it comes. Actually, winters here are about like weather conditions in the spring and fall up in north country. So bring some warm, layered-type clothing and leather gloves, and you'll be prepared for some great collecting times!

Are the shapes of these crystals natural? It looks like they have saw marks on them!

Yes, the shapes of the crystals are natural. All quartz crystals have six sides due to their internal structure. They are not faceted. Some crystals grow with more pointy terminations on the end than others. The shape of the point depends on the conditions in the earth when the crystals were forming. The lines that look like saw marks are striations, almost always present on the sides. They are growth features, much like the ridges on your fingernails.

A very few specimens you see in the shops have been polished, which makes the surface appear smoother and more lustrous than a plain crystal. For the most part, dealers spend enough time just cleaning and trimming the crystals. If they are going to cut the crystals, they would be more likely to do faceting or lapidary work with them.

What's the crystal festival?

The World Championship Quartz Dig is sponsored by the Mount Ida Chamber of Commerce in association with local businesses as part of the annual Quartz, Quiltz and Craftz festival. Several local mines participate, and diggers come and spend three days digging competitively for the best singles and clusters. Judges score each day's find. At the end of the dig winners with the highest points share $1500 in cash prizes. It is held the second weekend in October. You can get more information from the Mount Ida Chamber of Commerce. The address is given in the back of this book.

Is quartz associated with gold?

In some parts of the country it is. In Arkansas it is not, because there is no gold here in recoverable quantities. There have been two major gold scams in Arkansas, and many people have prospected for gold with no success.

What are crystals worth?

Museum quality doesn't have to be big or terribly expensive. Several factors enter into determining the "value" of any mineral specimen. Please remember "beauty is in the eye of the beholder." And that "beauty" may be precisely what makes one crystal more pricey than another. The overall appearance counts toward beauty... we may not be able to define it well, but we certainly know it when we see it.

In gemstones, the 4 "C's" are the determining factors when grading a stone, and to some extent those C's extend to crystals. Clarity, color, carat, and cut are the buzz words for gems. "Cut" is not applicable here, but maybe we could substitute "condition", which would include flaws or breakage. Collectability is also a factor.

CLARITY How transparent is the crystal to light? If milky

from tip to base, then the value is less than if the tip or termination is transparent. If the entire crystal is clear from tip to base, then it has more value than if partly milky. "Crystal clear" either came from quartz crystal or from blown glass crystal ware, but you get the idea.

COLOR Aside from its clarity making the crystal more valuable, there are certain colors you will find in quartz. In this case, color relates to being a more rare specimen, and if it is aesthetically pleasing, it will be more valuable.

Generally speaking of Arkansas quartz, the lack of color, rather than being milky, makes the value of the crystal increase.

SIZE (OR CARAT) In both common and rare minerals, size usually does matter. Given all other factors equal, bigger is more expensive. You would expect to pay more for a coffee table-sized quartz cluster than for one you can easily hold in your hand! Many crystals are priced per pound.

CONDITION may include these other factors:

INTERNAL FLAWS If the crystal contains fractures or cracks, then it is not as valuable as if it is internally flawless.

EXTERNAL BREAKAGE The key question is: Is it natural and rehealed breakage or did the specimen get damaged due to mishandling by whoever collected and/or cleaned it? Externally flawless specimens from Arkansas are relatively scarce, due to both reasons mentioned above. Anytime you see quartz piled up on a table, it will be difficult to find an undamaged specimen. Very small chips or dings (as they are called) can be tolerated if the specimen is otherwise aesthetically interesting. Natural breakage is due to movements of the earth along fractures and faults. Many of the early-formed quartz veins were in fractures, which because they were in a weak spot in the ground, were reopened by later earth movement. Then more quartz formed.

Unusual forms and highly distorted forms often come from pockets formed by multiple fracture episodes. Rarely, faulted crystals are recovered, sometimes healed back together, but with an offset of the two pieces. Sometimes we find two halves that fit back together. Anyway, natural breakage surfaces almost always have some crystal regrowth on the "broken" faces, whereas miner's breakage appears as small to large sites of conchoidal fracture. Miner's breakage reduces the value of the specimen more than natural breakage, especially if it is on the tip or termination of any crystal. Needless to say, a specimen with no breakage is more valuable than one with any type of breakage. Sometimes, if the broken crystal or damaged portion of the specimen is removable, an individual specimen's value may be greatly increased by some careful trimming.

LUSTER Luster is the amount of light that reflects from crystal faces. Specimens with high luster are more valued than those with lower luster. You want a specimen that is shiny and sparkly, not dull and drab in appearance.

MATRIX Does the specimen have matrix attached or not? If so, then the less the matrix the higher the value. Matrix adds weight and an opaqueness to the overall appearance of a quartz specimen. Sometimes matrix actually may add value if it allows the specimen to sit nicely without having to use a stand. If there is no matrix present, does the back appear to have been broken naturally or has it been broken by the miner from the host rock? To me, natural breaks are not really damage. Sometimes during late earth movement, entire sections of vein or linings of crystal pockets have been broken loose from the wall rock. These are fun to collect as they are often loose in the pocket and are just easily removed! They often appear as plates of crystals.

AESTHETICS Is there something about the specimen that really "grabs" you? Or do you skip right over it when looking through a box of specimens? After more than 30 years of looking at quartz, you might think that I would get bored. Sometimes I do, but you never know what you might see that will "turn you on" to a particular piece. Artists know the quality of aesthetics in mineral specimens better than most collectors. They have an "eye" for what looks attractive. I look for unusual features on any specimen, such as crystal form, crystal shape, clarity, position of the various crystals, display potential to show a particular feature, and if there are any other interesting associated minerals present, either on the surface or as inclusions. Does the specimen sit up on its own or does it need a stand for it to display properly? It will be worth considerably more if it does not need a stand. Is it an unusually attractive arrangement of crystals or are the crystals unusually large for the overall size of the piece? A sample consisting of a couple of 6-inch long crystals attractively extending from the matrix of a piece coated with 1-inch long crystals is certainly more valuable than the same specimen without the longer crystals. A cluster with points that create a center of interest is very pleasing to look at.

SCARCITY This is a rather difficult characteristic to get a handle on. We do not normally think of quartz crystal as being scarce, but certain forms, habits, and inclusions are less common than others. Does the specimen possess some unique or special characteristic, such as fluid inclusions, phantoms, or tabular shape, that increases its value? The presence of any one of these may increase the value of an otherwise mundane specimen. About 1995, a pocket of quartz crystal was recovered from a mine in Saline County that had shiny small galena inclusions, making otherwise typical quartz specimens rather unusual. Most of these specimens made one of the Tucson satellite shows at very inflated prices. A few specimens sold at about 40% of the marked prices near the end of the show. I didn't buy even a thumbnail-size specimen because the price was too high. I still do not have a single example from this pocket in my collection, even though I was the geologist who identified the included mineral as galena. I think the original purchaser of the entire pocket thought the inclusions were native silver! Anyway, since this is the only pocket of galena-included quartz found in the past 30 years, it is certainly scarce and worth more than normal quartz. It is really up to the collector to decide if the seller is way out of line on his prices. After all, until you hand over your money, it is your decision to make.

JEWELRY POINTS bring top dollar. Jewelry points are usually slender clear lustrous points that have clean terminations on one end. Crystals range from 0.5 to 2.5 inches in length. It may take 100 to 300 of these crystals to weigh a pound. Prices range considerably ($30 to $300/pound) from dealer to dealer and so does quality. These points are used, often mounted in sterling silver or gold-plated findings, in necklaces, earrings, or pendants.

COLLECTOR'S VALUE To most beginning rockhounds, a specimen they personally collected has more intrinsic personal value than one they purchased, traded for, or had given to them. The reason is simple. After investing so much time and effort in finding, cleaning, and trimming a specimen, you develop an emotional attachment to it. That is why it is so hard to admit you really should not have bothered with it in the first place and it is really a piece of leaverite! I have known individuals who moved and carried truckloads of this stuff with them from place to place around the country. As a collector becomes more experienced, he/she soon realizes that it is difficult to collect a truly good quality specimen for several reasons. Therefore, the silver pick at any given rockshop may turn up a specimen which you may never have a chance to collect personally. I do appreciate the individual who wishes to put together a "self-collected" mineral set, but I hope they realize the limitations they put on the potential value of their collection by going down this road.

One individual, now deceased, who was in my local

club often talked about needing to dispose of his collection and selling it, but he could not take the time to get it appraised by anyone. After he passed away, I had the sad duty to tell his widow that although her husband had personally valued his collection highly, it really consisted of so much yard rock. I was surprised when she stated that she had suspected it all along and appreciated my honesty. The rocks now reside in several flower beds and she has two extra usable rooms in her home.

What about buying quartz?

Quartz crystal can be purchased as single pieces or in bulk as uncleaned mine-run material. Most of the dealers will have shops that include outside tables. It is not such a good idea to buy quartz while it is wet, as you can't always see the damage on it. If it is raining, or the pieces have just

The author on left, and Stu Schmidt show some mine run baskets that Stu is getting ready to clean and process.

been washed, you won't get a true picture of what the crystals will really look like. A friend of ours who is now deceased, Charles Milton, told me when he was a kid he used to pick up rocks in the creek because they were so pretty. He was always disappointed when he got them home because they had dried out and weren't shiny anymore!

Mine-run quartz, meaning baskets of quartz that have not been cleaned or worked on in any way, can be purchased from some dealers. Prices range from $25 on up, with discounts being significant for 100 baskets or more. Sonny Stanley often puts out recently mined material on screen tables to let the clay dry so he can come back and wash the dried clay off. If you get there after he washes and before he cleans the material in acid, he will sell it right off the table as is. If you want to buy some bulk material in baskets, he usually has baskets stacked in his storage building. You may look around and pick a likely basket or baskets, but no shuffling of specimens from one basket to another is allowed. In years past, baskets have sold for as little as $25 to as much as $75, depending on what he judges the quality of the overall stacked baskets to be.

A wrapper basket is a more expensive mine run lot that has the top layer of pieces wrapped to prevent damage.

When purchasing any quartz this way, it takes an educated eye to recognize if the price is reasonable or too high for the general quality you will get. The only way to educate yourself is to purchase a few baskets over time and clean the quartz yourself. If you aren't happy after doing this a few times, then you should realize that you are just not yet experienced enough at grading uncleaned specimen material. I have done this for several years and still occasionally get a basket that does not measure up to what I thought its potential was. But sometimes I get a single surprisingly good specimen worth more than the cost of the entire basket.

How do dealers price quartz?

The price of quartz varies with the experience of the person grading the material and the seller's knowledge of their market. You may be buying directly from the miner or from a dealer who resells minerals. Part of the price they charge depends on how much they had to pay for the material, how long they have had it, their intended market, and the overhead of running their business. It's a function of supply and demand or what the market will bear. Unlike a value on the stock market, or the gold standard, there is no set price that quartz sells for. If you are buying at a show, you might get a price break on the last day, when the dealer faces the prospect of packing the material to take it home!

Commercially speaking, dealers sell quartz by the pound either wholesale or retail. Uncleaned mine-run specimen material may cost from $4-$6 per pound. To pick off a table of this material with some of the clay washed off, may cost you $8-$10 per pound. Cleaned clear specimens in small sizes often cost between $10-$15 per pound. Aesthetic pieces in the same size range run from $25-$50 per pound. Then come "collector" specimens. These specimens generally have all the positive characteristics previously mentioned and are in some way thought to be aesthetically attractive. Pick carefully if you decide to purchase from this type of quartz, as we are talking around $100 per pound. I really put my "critical eye" to the test when I see specimens priced in this category.

I also must say something about wholesale versus retail. With some quartz dealers, this entire situation is a joke, but to others it is a very serious matter. You can tell some dealers think it's a joke if they do not ask for a tax number, but just say, "Yep, all our prices are wholesale."

When you look to buy one or two cleaned and prepared specimens don't expect to get a truly wholesale price, even though the dealer may tell you it is. They may actually reduce the price 50% if you pull out a verifiable tax number. But they still have to make their profit above the expenses of a mining contract, mining equipment and labor costs, and pay for time involved in cleaning and grading the material.

If you want to purchase 100 to 1000 baskets of uncleaned mine-run quartz, then you will see what the true wholesale price is. Generally for equal quality specimen material, the more you purchase from a dealer in a single lot or sale, the cheaper the unit price will be. Price will vary for the same quality material from dealer to dealer, so it pays to visit several shops before making major purchases.

What is crystal healing?

Used as an aid in meditation and holistic alternative medicine, the general category is called "healing". At the time of the Renaissance in Europe, people began a questioning search for answers to the world around them. The belief in the power of stones to guard, protect, and cure was so entrenched in the culture of that time that no investigations were deemed necessary to explain anything so obvious as the beneficial results of using crystals and other minerals and gemstones. As medicine and pharmacology began to grow, the use of stones for healing gradually fell into disuse in favor of modern cures and treatments.

The New Age movement of the 1980's brought a renewed interest in the power of stones, quartz crystals especially, to transmit and receive energy for one's spiritual well being and overall health.

Controversy over this topic comes from several directions. The scientific community doesn't recognize the effects because they are subjective and not objective. An even bigger storm comes from religious conflicts. Related to eastern mysticism, the spiritual users of crystals say the information coming from and about the crystals are "channeled" from a non-physical entity, or disembodied spirit.

Judeo-Christian religions cite references from Leviticus and Deuteronomy prohibiting these kinds of practices.

Certain crystals with different shapes or numbers of faces are reputed to have different mystical properties. The crystal ball is legendary in regard to foretelling the future and is often thought of in the hands of gypsies who give a vision for a price. The crystal ball gazers, or "skryers"—if you need a good scrabble word—would look into the shiny sphere for a long time until a vision born of eye fatigue and meditative spirit was produced.

Thousands of years ago, crystal balls were made in the orient by the hands of masters who spent decades rolling the balls in a progressively finer grit. Today crystal balls have been shaped by a machine, which rolls the roughly rounded sphere into a smoother and smoother ball, then it is polished.

Some rock shops carry glass and plastic spheres, which may be difficult to distinguish from a natural crystal ball. If you are unsure as to a sphere's genuine origin, spend some time looking into the sphere with both the naked eye and a good hand magnifier (10X preferably).

Some quartz will have cloudy patches or internal flaws, these can be seen with the unaided eye. If you see fluid swirl marks with the naked eye, the sphere is glass or plastic. Quartz never shows swirl marks. Under magnification, look for small spherical bubbles. If present, the sphere is not a natural material. Except for lava flows, nature almost never produces perfect spherical holes, especially not in rock crystal. These bubbles can occur in both glass and plastic spheres.

10. Facilities and Resources

AFTER YOU SPEND several hours, or days, digging crystals, you'll appreciate what you see at the rock shops. Called silver picking, purchasing a specimen is a way to get that special piece you couldn't find for yourself. If you are trying to find a particular kind of crystal, let the shop owner know. The people in the quartz business know each other and they may be able to direct you to a shop who has just what you need.

By and large the rock shops are relaxed places. The people-greeter is usually a dog. Sometimes the dogs look up at you with a wag, other times you have to step over them.

Rock Shops
Bob's Rock Shop 507 Hwy 71 North, DeQueen AR 71832
Boxer's Crystals 8643 Albert Pike Road, Royal, AR 71968 (501) 767-6979; toll free 1-877-767-6900. 13 miles west of Hot Springs on Hwy 270. Specializes in green inclusion quartz. Check with owner about digging at his mine. Niles Boxer. boxers@hotsprings.net
Brewer Mining 192 Diamond Mist Rd., Mt. Ida, AR 71957 (870) 867-4033 by appointment. Wholesale & retail metaphysical quartz, minerals, tools and books. They sell from their home and website, and can talk at length on crystal energies. The High Peak mine yields beautiful crystals. Patrick & Susan Brewer. www.a-metaphysical-mall.com
Caddo Antiques Box 669, Murfreesboro, AR 71958 (501) 285-2780. Minerals, fossils and artifacts. Sam Johnson. www.caddotc.com

Caddo River Rock and Gem PO Box 103, Caddo Gap, AR 71935 (870) 356-6369; 1-800-562-9576. Located on Hwys 27 & 8, 15 miles south of Mt. Ida. A small shop with a large assortment of inexpensive tourist minerals from all over the world. Bill Jones. www.caddoriverrock.n.gem.com

Coleman's Crystal Yard Highway 7 North, Jessieville, AR 71949 (501) 984-5328. Quartz, irradiated quartz, and gift items. Retail shop for the Miller Mountain mine. Jim Coleman.

Coleman's Rocks-R-Gems 1400 East Grand, Hot Springs, AR 71909 (501) 624-7280. Wholesale & retail quartz crystal, tourist items, assorted minerals & gifts. Ron Coleman.

Crystal Heaven 104 Sunrise Hills Drive, Mount Ida, AR 71957 (870) 867-4625. Four miles east of Mount Ida. Wholesale, retail, and dig your own crystals. Clear, white, silver, blue & black phantom crystals. Stuart Zove. www.crystalheaven.com

Crystal Hills Mining Co. Rt. 1 Box 926, Glenwood, AR 71943 (870) 356-4615 Wholesale and dig your own quartz. By appointment, call Tim Hill. www.eyesoftime.com/hillmine

Crystal Springs Mining, Inc. P.O. Box 40, 8649 Albert Pike Road, Royal, AR 71968 (501) 991-3557; toll free 1-877-836-0838. 13 miles west of Hot Springs. Wholesale and retail. Tinkerbells, Poona India Zeolites, Columbian and Arkansas Quartz. They once had a fee pay mine, but it is now closed. Thomas Nagin. csmc@hsnp.com

CTS Rocks & Minerals 1501 E. Grand, Hot Springs, AR 71901 (501) 622-6800. Hwy 70E at Gulpha Gorge exit. Arkansas minerals, minerals for advanced collectors, and gifts. Charles and Anne Steuart. ctsa@hsnp.com

Fantasy Forest 3273 Highway 71 South, Mena, AR 71953 (501) 394-5916. Richard Milligan.

Fiddler's Ridge Rock Shop 3752 Highway 270, Mount Ida, AR 71957 (870) 867-2127. Seven miles east of Mt. Ida. Dig your own crystals.Wholesale and retail quartz and gifts. Kathy & Jim Fecho. fecho@ipa.net

JD Rocks 1209 Hwy 270 West, Mt. Ida (870)867-5235. Tumbled stones, crystals, gems & jewelry. Jerry Boswell www.orbitworld.net/jdrocks

Judy's Crystals N Things (also known as Mountain Gems) P. O. Box 956, Mount Ida, AR 71957 (870) 867-2523. Just past the airport at Logan Gap Road and Highway 270. Wholesale and retail. They mine quartz from Sonny Stanley's mine on Fisher Mountain. Quartz, digger bars, metaphysical. Judy Morton and Ray McGrew. www.mtidachamber.com/judy

Js Bonanza Quartz and Cases P.O. Box 233, Mt. Ida, AR 71957 (870)867-6201. On Hwy 270 in town. Custom built display cases. Dan Frisbee.

Leatherhead Mining 5057 Hwy 270 West, Pencil Bluff, AR 71965 (870) 326-4871. Retail shop 8 miles west of Mt. Ida on Highway 270. Wholesale shop 12 miles west of Mt. Ida. Dig your own crystals by appointment. Tony Thacker. www.mtidachamber.com/leatherhead

Lucky Strike Rock Shop 2385 North Highway 7, Hot Springs, AR 71909 (501) 623-5423. Eight miles north of Hot Springs. Arkansas quartz, many with chlorite inclusions from the Gladstone mine. Robert Gossage.

McGregor & Watkins 2600 Airport Road, Hot Springs, AR 72913 (501) 760-1800. Arkansas and Mexican minerals. Shop was recently sold.
M&M Crystal Mining HC 68 Box 772, Plainview, AR 72857 (501) 440-2424. Wholesale quartz, by appointment. Ask for Michael or Mary.
Miller Mountain Crystals P.O. Box 21, Jessieville, AR 71949 (501) 984-5328. Dig your own crystals 9.1 miles west on 298 from Hwy 7. Wholesale & retail quartz. Bill & Lila Norris.
Miner's Camping & Rock Shop Rt. 1 Box 365, Murfreesboro, AR 71958 (870) 285-2722. Less than 1 mile south of the Crater of Diamonds on Ark Hwy 301; minerals. Chuck & Joyce Goodin. www.diamondark.com
Ouachita Mountain Rockhound Rockstop P.O. Box 113, Cove AR 71937 (870) 387-7718. Gem trees. Ken Headley. email: jhea81945@aol.com
Robins Mining Company Box 236, Mount Ida, AR 71957 (870) 867-2530. At the junction of Highway 270 and 27 south. Dig your own crystals. Wholesale & retail. One of the best quartz shops, very few tourist doodads. Mearl Robins. www.robinsmining.com
Ron Coleman Mining Little Blakely Road, Jessieville, AR 71909 (501) 984-5396. Dig your own crystals wholesale & retail; campground; Arkansas quartz & gifts. www.omtp.com/colemans
Stanley's Crystal Yard 513 S. Pine Street, Mount Ida, AR 71957 (870) 867-3556. Dig your own crystals. Wholesale and retail Arkansas quartz and a few other minerals. Sonny Stanley. www.mtidachamber.com/stanley
Starfire Quartz Crystal Mines 5403 Hwy. 270 East Mount Ida, AR 71957 (870) 867-2431. Dig your own crystals. Located at the Colonial Grocery and Motel on Highway 270 at Mt. Harbor Road. Charlie, Shirley and Chris Burch. Arkansas quartz. www.starfirecrystals.com
Sweet Surrender Crystal Mine 60 Mary's Eagle Trail, Mt Ida, AR 71957 (870) 867-2443. By appointment. Quartz, metaphysical crystals. A good selection of quartz from the different Mt. Ida mines; a small shop by the house. Stuart Schmitt. www.eyesoftime.com/sos
TNT Rocks & Gifts (formerly Crystal Pyramid) 2383 Highway 270 East, Mt Ida, AR 71957 (870) 867-5203 Wholesale and retail quartz, dig your own crystals. Darren and Sheila Harris. email: nika@alltel.com
Vapor Valley Rock & Gem 344 Central Avenue, Hot Springs, AR 71901 (501) 624-7814. Across from Bathhouse Row. Arkansas quartz, assorted minerals, indian art. Charles and Leah Holmes.
Wegner Enterprises, Inc. 82 Wegner Ranch Road, Norman, AR 71957 (870) 867-2309. Three miles south of Mt. Ida on Highway 27, then left on Owley Road. Dig your own crystals by reservation, groups of 10 or more only. Wholesale, retail. Quartz, blue phantoms, and assorted minerals. Richard Wegner. wegnercrystalmines.com
West's Crystal Mining Company, 2425 Highway 270 East, Mount Ida, AR 71957. (870) 867-3994; 1-800-457-9197. Four miles east of Mount Ida at Hurricane Grove. Large display with interesting quartz specimens. Wholesale and retail quartz. Kenneth & Brenda Manley.
Wright's Rock Shop 3612 Albert Pike Rd., Hot Springs, AR 71913 (501)

767-4800. World wide minerals, Arkansas' most complete rockshop. Chris Wright. www.wrightsrockshop.com

Arkansas State Park Campgrounds

Lake Ouachita State Park. 112 campsites (Preferred, Class A and Class B), picnic areas, a marina (with boat, motor and slip rentals; bait and supplies), swimming area and trails. A store and snack bar are located in the visitor center. Fully equipped cabins with kitchens overlook the lake. From Hot Springs, travel three miles west on U.S. 270, then go 12 miles north on Ark. 227 to the park. 5451 Mountain Pine Road, Mountain Pine, AR 71956. Park & Campsite reservation: (501)767-9366, Cabin reservations: 1-800-264-2441.

Lake Catherine State Park 70 Class A campsites including Preferred sites and Rent-A-Camp. Rental boats, a marina and launch ramp, pavilion, picnic sites, playground, laundry and trails. 17 fully equipped cabins. Between Malvern and Hot Springs. 1200 Catherine Park Road, Hot Springs, AR 71913. Park and Campsite reservations: (501) 844-4176 Cabins: 1-800-264-2422.

Hot Springs National Park Campground

Gulpha Gorge Campground and picnic area 47 Tent or self contained vehicle sites. No hookups. Central flush toilets, water, and trailer dump station. Camping limit 14 days per year, open year-round. User's fee charged for camping. One half mile northeast of Hot Springs off US 70B on the east side of town. Our cub scouts go crawdad fishing here.

U.S. Forest Service Campgrounds
Ouachita National Forest

A fee is charged to use some National Forest recreation areas. Fees vary based on the type of services provided at each area.
Caddo Ranger District, Glenwood AR 71943 (870) 356-4186.
Womble Ranger District, Mount Ida, AR 71957 (870) 867-2101.
Jessieville Ranger District, Jessieville, AR 71949 (501) 984-5313.
Winona Ranger district, Perryville, AR 72126, (501) 889-5176.
Mena Ranger District, Mena, AR 71953, (870) 394-2382

Albert Pike 46 RV/Tent sites with no hookups. 5 miles southwest of Glenwood on US 70, then west 13 miles on State 84 to Langley, then right on Highway 369 for 6.1 miles. Attractions: Natural pool in the Little Missouri River, swimming, fishing. Facilities: Central flush toilets, trailer dump station, warm showers, water. (Partially open year-round). User's fee charged. *Caddo Ranger District.*

Bard Springs 17 RV/Tent with no hookups. 19 miles west of Norman on State Highway 8 to Big Fork, then 10 miles south on gravel Forest Road

38. Attractions: small swimming area, hiking trails, fishing. Facilities: central flush toilets, water, picnic shelter (Partially open year-round). User's fee charged. *Caddo Ranger District.*

Big Brushy 11 RV/Tent with no hookups. 16 miles west of Mt. Ida on US 270. Attractions: Hiking and fishing. Facilities: Vault toilets, water. (Open Apr. - mid Nov.) *Womble Ranger District.*

Camp Clearfork Group Use Area 20 miles west of Hot Springs on US 270. Attractions: Swimming, fishing, hiking, canoeing and field sports. Facilities: 6 dormitory cabins, 3 staff cabins, dining hall with full kitchen facilities and recreation building. Central bathhouse with flush toilets and warm shower, ball field. (Some facilities are accessible to the physically challenged.) 30 day minimum advance reservation plus deposit. *Womble Ranger District.*

Charlton 69 RV/Tent sites with no hookups. Attractions: Located near Lake Ouachita. Swimming, mountain streams, tree identification interpretive trail. Hiking to Lake Ouachita. Outdoor theater programs in season. Facilities: Central flush toilets, cold showers, water, picnic sites/trailer dump station. Open mid May - mid Nov. Users fee charged. Some campsites available by reservation, call 1-800-CAMP. *Womble Ranger District.*

Crystal 9 RV/Tent with no hookups. 1 mile north of Norman on State 27, then turn east for 3 miles on gravel Forest Road 177. Attractions: swimming, interpretive trail on soil formation, scenic drives. Facilities: Flush toilets, water, picnic sites. (Open year-round). User's fee charged. *Caddo Ranger District.*

Dragover 8 RV/Tent with no hookups. 4 miles east of Sims on Hwy. 88 east, then south on County gravel road 97 for 2 miles. Attractions: Float fishing on Ouachita River, swimming, hiking. Facilities: Vault toilets, water, picnic sites, boat ramp. Open year-round. *Oden Ranger District.*

Fourche Mountain 5 RV/Tent with no hookups. 5 miles south of Rover on Arkansas Hwy. 27. Attraction: Hiking. Facilities: vault toilets, water, picnic sites. Open year-round. *Fourche Ranger District.*

Fulton Branch Float Camp 7 RV/Tent with no hookups. 1 mile northeast of Mount Ida on State 27, then north at sign for 6 miles on gravel road. Attractions: Float fishing on the Ouachita River, swimming, hiking. Womble Trail connects to Ouachita National Recreation Trail. Facilities: Vault toilets, water, picnic sites, boat ramp. Open year-round. *Womble Ranger District.*

Hickory Nut Mountain Picnic area 8 RV/Tent with no hookups. 24 miles west of Hot Springs on US 270, then turn north at sign for 6 miles on gravel Forest Road 47 Attractions: Panoramic view of Lake Ouachita. 1 mile to Lake Ouachita vista area. Facilities: vault toilets, picnic sites. Open year-round. *Womble Ranger District.*

Iron Springs 13 RV/Tent with no hookups. 26 miles north of Hot Springs on State 7. Attractions: Hiking trails and small wading area. Hunt Trail connects to Ouachita National Recreation Trail. Facilities: Vault toilets, water, picnic sites. Open year-round. *Jessieville Ranger District.*

Lake Sylvia 19 RV/Tent with electric. 9 miles south of Perryville on State 9, then west at sign for 4 miles on gravel State 324. Attractions: 14 acre lake, swimming beach, boating (no motors) fishing, hiking. Wildlife interpretive trail, and tree identification interpretive trail for handicapped. Outdoor theater programs in season. Facilities: Central flush toilets, water, hot showers, trailer dump stations. Open mid May - early Sept. User's fee charged. *Winona Ranger District.*

Little Missouri Falls RV/Tent with no hookups. Take Arkansas Hwy 84 west of Glenwood, AR; turn north on Arkansas Hwy 369 at Langley, AR for 6 miles; continue north 3 miles on road 73 to Forest Service Road 43; turn left (northwest) 4 miles to Forest Service Road 25; turn left (west) for 0.5 mile to Forest Service Road 539. Facilities: Flush toilets, water, picnic sites. (Open year-round). User's fee. *Caddo Ranger District.*

River Bluff 7 RV/Tent with no hookups. 1 mile northeast of Mount Ida on State 27, then north at sign for 5 miles on gravel road. Watch for signs. Attractions: Float fishing on Ouachita River, swimming, hiking. Womble Trail connects to Ouachita National Recreation Trail. Facilities: Vault toilets, water, picnic sites, boat ramp. (Open year-round). *Womble Ranger District.*

Rocky Shoals Float Camp 7 RV/Tent with no hookups. 2 miles southeast of Pencil Bluff on US 270. Attractions: Base camp for float fishing on Ouachita River, swimming, hiking. Womble Trail connects to Ouachita National Recreation Trail. Facilities: Vault toilets, water, picnic sites, boat ramp. Open year-round. *Womble Ranger District.*

Shady Lake 96 RV/Tent with no hookups. 5 miles southwest of Glenwood on US 70; west 23 miles on State 84 to Athens, then north at sign for 5 miles on Forest Road 38. Attractions: 25 acre lake, swimming beach, fishing and hiking. Outdoor theater programs in season, tree identification interpretive trail. Facilities: Central flush toilets, water, hot showers, trailer dump station, picnic sites, boat dock (no motors). (Partially open year-round). User's fee charged. *Mena Ranger District.*

Corps of Engineers / Lake Ouachita Campsites

Reserve sites in advance from the New York base toll-free 1-877-444-6777. Phone answered 8am-midnight Eastern Time in summer, 10am to 7pm the rest of the year. Walk in sites also available.

Big Fir 17 RV/Tent with no hookups. 6 miles northeast of Mount Ida on State 27, 6 miles east on 188, then 4 miles east on gravel road. Vault toilets, water, boat ramp. Open year round.

Brady Mountain 57 RV/Tent with electricity, 17 tent sites with handicapped access. 10 miles west of Hot Spring on US 270, then 7 miles north on access road. Flush toilets, water, hot showers, boat ramps, dump station, nature trail, restaurant, marina, store, playground, scenic overlook, scuba shop, pavilion, lodge. Open year round. Fee area.

Buckville 6 RV/Tent with no hookups. 10 miles northeast of Hot Springs on highway 7, west 18 miles on 298, then 7 miles south on Buckville road. Vault toilets, water, boat ramp, beach. Open year round.

Crystal Springs 53 RV/Tent sites with water & electricity, 21 tent sites with handicapped access. 15 miles west of Hot Springs on 270, then 2 miles north on access road. Flush toilets, water, hot showers, boat ramp, dump station, picnic sites, 2 pavilions, day use swimming beach with change house and playground, restaurant, marina, store, cottages. Open year round. Fee area.

Denby Point 58 RV/Tent sites with electric, 9 tent sites. 8 miles east of Mount Ida on US 270, then 1 mile north on access road. Flush and chemical toilets, water, hot showers, dump, 4 picnic sites, amphitheater, nature trails, restaurant, marina, 2 group camping areas (6-site and 7-site reservations accepted) swimming beach, lodge. Open year round. Fee area.

Highway 27 19 RV/Tent with no hookups. 9 miles northeast of Mount Ida on Highway 27. Flush toilets, water, boat ramp, dump station, beach, restaurant, marina, ball field, horseshoe pits. Open year round. Fee area.

Irons Fork 5 RV/Tent with no hookups. 8 miles east of Story on 298, then 1.5 miles on access road. Vault toilets, water, boat ramp. Open year round.

Joplin (Mountain Harbor) 65 RV/Tent with no hookups. 11 miles east of Mount Ida on US 270, then 2 miles north on access road. Flush toilets, water, showers, boat ramp, dump station, picnic sites, swimming beach, restaurant, marina, lodge. Open year round. Fee charged April - Sept.

Lena Landing 10 RV/Tent with no hookups. 12 miles west of Blue Springs on Highway 198, then one mile south on Navy Landing Road. Flush toilets, water, boat ramp, dump station, restaurant, marina, convenience store, scuba shop. Partly open year round. Fee charged Apr through Sept.

Little Fir 25 RV/Tent with no hookups. 6 miles northeast of Mount Ida on highway 27 then 9 miles east on 188. Flush toilets, water, boat ramp, dump station, marina. Open year round. Fee charged Apr through Sept.

Stephens Park 9 RV/Tent with electricity. At Blakely Mountain Dam, 1 mile west of Mountain Pine on Highway 227. Flush toilets, water, boat ramp, pavilion, 8 picnic sites, playground. Open year round. Fee area.

Thompkins Bend (Shangri-La) 47 RV/T with electricity, 16 RV/Tent with no hookups. 14 tent sites. 10 miles east of Mount Ida on US 270, then 3 miles north on access road. Flush toilets, water, hot showers, boat ramp, dump station, amphitheater, restaurant, marina, cabins. Open year round. Fee area.

Twin Creek 15 RV/Tent with no hookups. 8 miles east of Mount Ida on US 270, then 1 mile north on access road. Flush toilets, water, boat ramp, dump station, swimming beach. Open year round.

Privately Owned Campgrounds
Hot Springs
All Season Mobile & RV Park 10 sites. Highway 7 south, 4.5 miles south of

the mall, 6507 Central Ave, Hot Springs 71913. (501) 525-1248.
Camp Lake Hamilton 40 sites Highway 7 south, 6191 Central Avenue, Hot Springs, AR 71913. (501) 525-8204.
Hot Springs KOA 112 sites. US 70 east of Hot Springs, exit 4, 838 McClendon Road, Hot Springs, AR 71901. (501) 624-5912.
Lakeside Trailer Park & Cottages 20 sites. Highway 7 south, 451 Lakeland Drive, Hot Springs, AR 71913. (501) 525-8878.
Rovin Ramblers Mobile & RV Park 15 sites. 15 miles SW of Hot Springs, Hwy 70. Rt. 1 Box 180 Bonnerdale, AR 71933. (501) 356-4412.
Youngs Lakeshore RV Resort (501)767-7946 Lakeshore Drive at McLeod.
Hot Springs RV Park 2345 East Grand, Hot Springs, AR 71901. (501) 623-5559.

Jessieville

John Teal's RV Park 10 full hookups with cable. On Highway 289 at Highway 7, behind the Fina Station.
Crystal Ridge RV Park On the grounds of Ron Coleman Mining, Inc. 24 sites with water and electric, showerhouse, washateria, dump station. Walking distance to fee-pay mining area. (501) 984-5396

Mount Ida

Marilyn's RV 11 sites with water and electricity. Highway 270 at the antique shop, by the airport.

Lodging

Mount Ida Area

Crystal Inn Motel 15 units P.O. Box 816, Mount Ida, AR 71957 (870) 867-2643. No Pets. Six miles east of Mount Ida on Hwy. 270 near the airport.
Colonial Motel 11 units 5403 Hwy 270 East, Mount Ida, AR 71957 (870) 867-2431. Pets extra. Eleven miles east of Mount Ida on Hwy. 270 crystals@ipa.net
Crystal Springs Resort 1647 N Crystal Springs, Royal AR 71968 (501) 991-3361 Between Hot Springs and Mount Ida.
Gap Creek Falls Motel HC 67 Box 48C, Mount Ida, AR 71957 (870) 867-2783 12 miles east of Mount Ida on Hwy 270. Two newly remodeled kitchenettes, with full bed in each. Air conditioned. Convenience stores nearby. Five minutes from Lake Ouachita.
Mount Ida Motel 17 units 222 Hwy 270 East, Mount Ida, AR 71957 (870) 867-3456 Small pets ok. East of the Court House Square in Mount Ida.
Mountain Harbor 75 units P.O. Box 1268, Mount Ida, AR 71957 (870) 867-2191 or (870) 867-1200; toll free 1-800-832-2276. Extra charge for pets. Full Marina, water toys. Eleven miles east of Mount Ida on Hwy. 270 On Lake Ouachita. mtharbor@ipa.net

Royal Oak Inn 15 units 936 Hwy 270 East, Mount Ida, AR 71957 (870) 867-2169. No Pets. Half mile east of Mount Ida.

Shady Rest Motel & Storage 5191 Hwy 270 East, Mount Ida, AR 71957 (870) 867-2455. No Pets. Housekeeping units, storage units 10.5 miles east of Mount Ida on Hwy 270.

Shangri-La Resort 28 units 987 Shangri La Drive, Mount Ida 71957 (870) 867-2011. Extra charge for pets. Full Marina and restaurant. Located ten miles east of Mount Ida on Hwy 270 on Lake Ouachita.

Highway 27 Fishing Village 12 units Family owned and operated.   214 Fishing Village Rd., Story, AR 71970 (870) 867-2211. Pets OK. Full Marina. Eight miles north of Mount Ida on Hwy. 27 On Lake Ouachita.

Lake Ouachita Shores Resort 20 units   359 Lake Ouachita Shores Parkway, Mount Ida, AR   71957   (870) 867-3651. Extra charge for pets. Full marina and gift shop. Marina number (870) 867-4646. Nine miles east of Mount Ida on Hwy 270 on Lake Ouachita.

Arrowhead Cabin & Canoe Rentals 5 Cabins, 2 bedroom house, 2 bed room mobile home. (870) 356-2944, (800) 538-6578.   Located on the Caddo River near Caddo Gap - turn at the junction of Hwys 8 and 240. Cross over the river and take the third drive on your right.

High Shoals Campground & Cabins   P.O. Box 523, Mount Ida, AR   71957 (870) 867-3761. Primitive cabins, self-contained modern 26' campers, camping, showers, canoes, Located on the Ouachita River.   Extra charge for pets.

Ouachita Mountain Outdoor Center P.O. Box 65, Pencil Bluff, AR 71965 (870) 326-5517, (800) 748-3718. Riverside cabin accommodates 12 people. Located on the Ouachita River. Eight miles west of Mount Ida on Hwy. 270. mdavis@ouachita-outdoor.com

Ouachita River Haven Resort - Cabins HC 64, Box 80, Pencil Bluff 71965 (870) 326-4941, 122 Ouachita River Haven Rd. Located on the Ouachita River East of Pencil Bluff on Hwy 270.   rebekah@ouachitahaven.com

River's Edge Bed & Breakfast 5 units, cabins HC 65, Box 5, Caddo Gap 71935 (870) 356-4864. Located on the Caddo River at Caddo Gap. Highways 8 and 240 West, 15 miles south of Mount Ida and   35 miles SW of Hot Springs National Park. riversbb@ipa.net

River View Cabins & Canoes   HC 69 Box 64, Oden, AR 71961(870) 326-4630 or (888) 547-1146. On the Ouachita River. Cabins, canoes, fishing, swimming and horseback rides. Located just south of Oden, AR.

Top Gunn Striper Guide Service & Lodging - Cabin Housekeeping Units 5148 Hwy 270 East, Mount Ida, AR 71957-9766 (870) 867-4086. Extra charge for pets. Located 10.5 miles east of Mount Ida on Hwy. 270. Five minutes from Lake Ouachita.

Hot Springs has all the major chains plus many other motels too numerous to list. Being a convention city, it may be wise to make reservations in advance. www.hotsprings.org

Glossary

asterism - property of a mineral that displays a star when viewed in reflected light
bladed - a shape term used to describe a mineral that forms an elongate flattened crystal
cabochon - rounded, dome-cut stone
candle - quartz crystal with a ratio of 6:1 (or greater) length to width
carat - unit weight used for gemstones; five carats in a gram
cavitation - the growth and collapse of bubbles in a fluid
chemistry - the study of elements and chemicals; also reactions
clay - an extremely fine-grained mineral composed of hydrous aluminum silicate
cleavage - the breaking of a mineral along its crystallographic planes, thus reflecting internal crystal structure
crystalline - having an internally ordered arrangement of atoms and/or molecules
Cretaceous - a geologic time period of 144 to 66 million years before present
cryptocrystalline - a crystalline texture that is so fine-grained that individual grains cannot be seen with an optical microscope

crystallographic axis - one of several imaginary lines that pass through the center of a crystal which are used as a reference
crystallography - the study of crystals, including their growth, structure, physical properties, and classification by form
deposit - mineral matter accumulated by natural processes
detritus (detrital) - a product of disintegration and/or wearing away of a rock
Devonian - a geologic time period ranging from about 410 to 360 million years ago
drusy - refers to the interior of a cavity encrusted with many small projecting crystals
element - any of more that 100 fundamental substances that consist of atoms of only one kind and that singly or in combination make all matter
enantiomorphism - the characteristic of two crystals being mirror images of each other
equidimensional - same distance in all directions; blocky
erosion - the general processes whereby the materials composing the crust are removed by natural processes
euhedral - a shape characterized by the presences of faces
exsolve - unmixing, usually in the solid state
fault - a rock fracture along which displacement occurred
fossil - prehistoric plant or animal remains preserved in rock
fracture - breakage in no particular direction
geologist - scientist who studies the earth
geology - study of the planet earth, especially its solid parts
geophysical - relating to the study of the physical properties of the earth
habit - the characteristic crystal form or combination of forms, including characteristic irregularities
hardness - the resistance of a mineral to scratching; in this text, based on Mohs' scale
hexagonal - a crystal system which has four crystallo-

graphic axes (a1, a2, a3, c) in which a1 = a2 = a3 ≠ c in length and all the a axes are at 60° to each other in the same plane, and c is 90° to that plane

hydrothermal - of or pertaining to heated water; in particular, minerals formed from a hot water solution

in situ - latin words meaning "in place"

inclusion - a particle or mineral encased in another mineral

infilling - a process of deposition

irradiated - bombarded by atomic energy

lapidary - a person who cuts and polishes stones

lime - calcium oxide; agricultural lime - calcium carbonate, limestone

luster - the reflection of light from a surface of a mineral; described by its color and intensity

magmatic - related to molten rock

matrix - the host rock in which a crystallized mineral is embedded in

microcrystalline - crystallized on a microscopic scale

mineral species - same as mineral

mineralogy - the study of minerals

Mississippian - a geologic time period ranging from about 360 to 320 million years ago

morphology - form

muriatic acid - hydrochloric acid, HCl

nucleation - process by which crystals begin to form

Ordovician - a geologic time period ranging from about 500 to 440 million years ago

orientation - in describing crystal form and symmetry, the placing of the crystal so its crystallographic axes are in the conventional position

oscillator - a radio-frequency generator

oxalic acid - a poisonous acid $H_2C_2O_4$ that occurs in various plants and is used as a bleaching or cleaning agent

Paleozoic - a geologic time period ranging from about 500

to 290 million years ago

Pennsylvanian - a geologic time period ranging from about 320 to 286 million years ago

phantom - a mineral with inclusions that displays an earlier crystal form

plate tectonics - global forces based on a model of crustal plates that "float" and move on viscous underlying rock

pocket - a mineralized cavity; same as a vug

polarized - said of light that vibrates in a single plane

prism - a crystal form open at both ends and consisting of faces having parallel edges

prismatic - referring to a mineral or crystal whose length is 1.5 to 3 times its width

pyramid - an open crystal form consisting of nonparallel faces that meet in a point

rhombohedral - a shape that is an oblique, equilateral parallelogram

rhyolitic lava - extrusive igneous rock, chemically equivalent to a granite

sandstone - a rock composed of sand-sized particles, usually quartz grains

sediment - fragmental soil material originating by the weathering of rocks and which is transported

sedimentary rock - rock formed by the transportation, deposition, and lithification of sediment

shale - a rock composed of very fine-grained sediment, often clay

silica - any form of the chemical SiO_2

silicified - replaced by some form of silica

silt - a fine-grained sediment

slate - metamorphosed shale

specific gravity - a number indicating the number of times heavier a body of any volume is than an equal volume of water; usually measured at room temperature and pressure

streak, streak test - refers to the color of a powdered mineral, usually on a white unglazed tile

striations - parallel lines on the surface of a crystal or cleavage plane

subparallel - almost parallel. In quartz, refers to growth relationship of the c axes of many crystals together

subduction zone - region where a crustal block descends relative to another crustal block.

tabular - describes a crystal form having two prominent parallel faces that give the crystal a broad, flat appearance

tectonic - said of mountain building forces

termination - point; literally, the end

tetrahedron - a four-faced geometric body, each face of which is a triangle

translucent - said of a mineral that is capable of transmitting light, but is not transparent

tribolumenescence - luminescence caused by friction

trigonal - a subclass of the hexagonal crystal system characterized by three-fold symmetry

uniaxial - said of a crystal having only one optic axis

uplift - a tectonically high area of the earth's crust, raised up by natural forces

volcanic vent - an opening in the earth's crust through which molten or hot rock and steam issue.

vein - a mineral filling in a fracture, often in tabular or sheetlike form

vug - same as pocket

Resources

Rockhounding Arkansas *The companion website to this book.* In depth information about collecting Arkansas minerals, managing a collection and more. For pebble puppies to advanced collectors. http://rockhoundingar.com

Arkansas Department of Parks and Tourism One Capitol Mall, Little Rock, AR 72201 (501) 682-7777. www.1800natural.com/

Arkansas Geological Commission 3815 W. Roosevelt Road, Little Rock, AR 72204 (501) 296-1877. www.state.ar.us/agc/agc.htm Source of publications and maps.

Elderhostel A nonprofit organization dedicated to educational needs of people 55 and over. Usually two crystal courses are offered each year in Hot Springs. The collecting field trip is preceded by study of crystals in the classroom. 800-895-0727 for free catalog www.elderhostel.org

Hot Springs, Arkansas Convention and Visitors Bureau 1-800-543-2284 134. Convention Blvd., Hot Springs, AR 71901. Local tourist information 321-2277. www.hotsprings.org

Mount Ida Chamber of Commerce PO Box 6, Mount Ida, AR 71957. (870) 867-2723. www.mtidachamber.com

Ouachita National Forest P.O. Box 1270, Hot Springs, AR 71902 (501) 321-5202. www.fs.fed.us/oonf/minerals/welcome.htm

University of Arkansas at Fayetteville Department of Geology 118 Ozark Hall University of Arkansas Fayetteville, AR 72701. (501) 575-3355. www.uark.edu/depts/geology/

University of Arkansas at Little Rock Earth Science Department, Fribourgh Hall Room 307 University of Arkansas at Little Rock, 2801 South University Ave., Little Rock, AR 72204 (501) 569-3546. www.ualr.edu/~ersc/index.html

About the authors

This husband and wife team have put together a popular website called Rockhounding Arkansas. Darcy Howard is a commercial artist and scientific illustrator who owns the Arts and Ideas Studio. She has illustrated college textbooks of mineralogy as well as other publications. Mike Howard is a geologist working for the Arkansas Geological Commission. An avid mineral collector himself, he has written many articles for collectors as well as the scientific community. Several of his articles have appeared in *Rocks and Minerals* magazine.